SECOND EDITION

PSPICE
and
MATLAB
for Electronics

An Integrated Approach

VLSI CIRCUITS SERIES
Series Editor: Wai-Kai Chen

PSPICE and MATLAB for Electronics: An Integrated Approach,
Second Edition
John Okyere Attia

Analog VLSI Design Automation
Sina Balkir, Günhan Dündar, and A. Selçuk Ögrenci

VLSI Design
M. Michael Vai

SECOND EDITION

PSPICE
and
MATLAB
for Electronics
An Integrated Approach

JOHN OKYERE ATTIA

CRC Press
Taylor & Francis Group
Boca Raton London New York

CRC Press is an imprint of the
Taylor & Francis Group, an **informa** business

CRC Press
Taylor & Francis Group
6000 Broken Sound Parkway NW, Suite 300
Boca Raton, FL 33487-2742

First issued in paperback 2018

© 2010 by Taylor and Francis Group, LLC
CRC Press is an imprint of Taylor & Francis Group, an Informa business

No claim to original U.S. Government works

ISBN-13: 978-1-4200-8658-4 (hbk)
ISBN-13: 978-1-138-37274-0 (pbk)

Library of Congress Cataloging-in-Publication Data

Attia, John Okyere.
   PSPICE and MATLAB for electronics : an integrated approach / John Okyere Attia.
-- 2nd ed.
      p. cm. -- (VLSI circuits series)
   Includes bibliographical references and index.
   ISBN 978-1-4200-8658-4 (hardcover : alk. paper)
      1. Integrated circuits, Very large scale integration--Design and construction--Data processing. 2. PSpice. 3. MATLAB. 4. Electronic circuit design--Data processing. I. Title. II. Series.

TK7874.75.A88 2010
621.39'50285--dc22   2010017180

**Visit the Taylor & Francis Web site at
http://www.taylorandfrancis.com**

**and the CRC Press Web site at
http://www.crcpress.com**

Dedicated to my Parents

for

their unfailing love and encouragement

Contents

Part II

List of Solved Examples

Preface

SPICE is one of the industry's standard software for circuit simulation. It can perform DC, AC, transient, Fourier, and Monte Carlo analysis. In addition, SPICE has device models incorporated into its package. There is an extensive library of device models available that a SPICE user can use for simulation and design. PSPICE, a SPICE package by Cadence Design, has an analog behavioral model facility that allows modeling of analog circuit functions by using mathematical equations, tables, and transfer functions. The above features of PSPICE are unmatched by any other scientific packages.

MATLAB® is primarily a tool for matrix computations. It has numerous functions for data processing and analysis. In addition, MATLAB has a rich set of plotting capabilities, which is integrated into the MATLAB package. Since MATLAB is also a programming environment, a user can extend the MATLAB functional capabilities by writing new modules (m_files).

This book uses the strong features of PSPICE and the powerful functions of MATLAB for electronic circuit analysis. PSPICE can be used to perform DC, AC, transient, Fourier, temperature, and Monte Carlo analysis of electronic circuits with device models and subsystem subcircuits. Then, MATLAB can be used to perform calculations of device parameters, curve fitting, numerical integration, numerical differentiation, statistical analysis, and two-dimensional and three-dimensional plots.

PSPICE has the postprocessor package, PROBE, which can be used for plotting PSPICE results. In addition, PROBE has built-in functions that can be used to do simple signal processing. However, the PROBE functions are extremely limited compared to those of MATLAB.

The goals in writing this book were to provide the reader with an introduction to PSPICE; to provide the reader with a simple, easy, hands-on introduction to MATLAB; and to demonstrate the combined power of PSPICE and MATLAB for solving electronics problems.

This book is unique. The book covers the introduction to both MATLAB and PSPICE. In addition, the book integrates the strong features of PSPICE and the powerful functions of MATLAB for problem solving in electronics.

Audience

This book can be used by students, professional engineers, and technicians. The first part is a basic introduction to the PSPICE software program. The second part of the book can be used as a primer to MATLAB®. This will be useful to all students and professionals who want a basic introduction to MATLAB. Part three is for electrical and electrical engineering technology students and professionals who want to use both PSPICE and MATLAB to explore the characteristics of semiconductor devices and to apply the two software packages for analysis and electronic circuits and systems.

Organization

The book is divided into three parts: Part I (Chapters 1, 2, and 3) is an introduction to PSPICE. ORCAD schematics are introduced in Chapter 1. The basic PSPICE commands are discussed in Chapter 2. The advanced features of PSPICE are covered in Chapter 3. The chapters have several examples to illustrate the application of PSPICE in electronics circuit analysis.

Part II (Chapters 4 and 5) is an introduction to MATLAB®. Circuit analysis and electronic applications using MATLAB are explored. It is recommended that the reader work through and experiment with the examples at the computer while reading Chapters 1 through 5. The hands-on approach is one of the best ways of learning PSPICE and MATLAB.

Part III includes Chapters 6, 7, and 8. The topics discussed in this part are diodes, operational amplifiers, and transistor circuits. The application of PSPICE and MATLAB for problem solving in electronics is discussed. Extensive examples showing the combined power of PSPICE and MATLAB for solving problems in electronics are presented. Each chapter has its own bibliography and problems.

In this second edition, ORCAD schematic capture is introduced in Chapter 1. The schematic capture and the PSPICE text programming approaches to circuit simulation are integrated in this edition. The steps for drawing and simulating circuits using the ORCAD schematic capture are summarized in "boxes" allowing the reader to easily use the ORCAD capture package. The coverage of MATLAB is updated with the addition of several MATLAB topics. Additional examples have been added. In addition, several problems have been added to each chapter. Furthermore, the bibliography at the end of each chapter has been revised and updated.

MATLAB® is a registered trademark of The MathWorks, Inc. For product information, please contact:

The MathWorks, Inc.
3 Apple Hill Drive
Natick, MA 01760-2098 USA
Tel: 508-647-7000
Fax: 508-647-7001
E-mail: info@mathworks.com
Web: www.mathworks.com

Acknowledgments

I am grateful to Monica Bibbs, Julian Farquharson, and Rodrigo Lozano for helping me to complete the first edition of this book. Special thanks to Nora Konopka, acquisitions editor at Taylor & Francis for her interest in this book. I thank Jill Jurgensen for managing the production aspects of this work.

Author

Dr. John Okyere Attia is professor and head of the electrical and computer engineering department at Prairie View A&M University in Texas. He has been teaching graduate and undergraduate courses in electrical and computer engineering in the field of electronics, circuit analysis, instrumentation systems, digital signal processing, and VLSI design for the past 28 years.

Dr. Attia earned his PhD in electrical engineering from the University of Houston, Texas; his MS from the University of Toronto, Canada; and his BS from Kwame Nkrumah University of Science and Technology, Ghana. In addition, he worked briefly at AT&T Bell Laboratories and 3M.

Dr. Attia has written over 65 publications. He is the author of the CRC Press publication, *Electronics and Circuits Analysis Using MATLAB®, 2nd Edition*. His research interests include innovative electronic circuit designs for radiation environment, signal processing, and radiation testing.

Dr. Attia has twice received outstanding teaching awards. He is a member of the Sigma Xi, Tau Beta Pi, Kappa Alpha Kappa, and Eta Kappa Nu. Dr. Attia is a registered professional engineer in Texas.

Part I

1

ORCAD PSPICE Capture Fundamentals

1.1 Introduction

SPICE (Simulated Program with Integrated Circuit Emphasis) is one of the industry standards software for circuit simulation. It can be used among other circuit analysis to perform alternating current, direct current, Fourier, and Monte-Carlo analysis. SPICE continues to be the standard for analog circuit simulation for the electronics industry over the past decades. There are several SPICE-derived simulation packages. Among these are ORCAD, PSPICE, Meta-software HSPICE, and Intusoft IS-SPICE.

PSPICE has additional features as compared to classical SPICE. Among some of the useful features are:

1. PSPICE has a post-processor program, PROBE, which can be used for interactive graphical display of simulation results.

2. Current flowing through an inductor, capacitor, resistor, can be easily obtained without inserting a current monitor in series with the passive elements.

3. PSPICE has analog behavioral model facility that allows modeling of analog circuit functions by using mathematical equations, tables, and transfer functions.

4. PSPICE does not distinguish between uppercase or lowercase character. In SPICE, all characters in the source file must be uppercase. (For example, rab and RAB are considered equivalent in PSPICE.)

1.2 PSPICE Schematics

1.2.1 Starting ORCAD Capture

The PSPICE discussed in this book runs under the Windows Operating System. Examples and instructions in this book are based on PSPICE

ORCAD Family Release 9.2 Lite Edition, provided by Cadence Design Systems. If the ORCAD PSPICE is installed on your computer, you can start the program by clicking on the "Start" icon, drag the cursor to "All Programs," then to "ORCAD Family Release 9.2" program, and then to "ORCAD Capture."

For a circuit to be drawn by PSPICE, you must (i) create the circuit, (ii) simulate it and (iii) print or plot the results. To create the circuit, select "**File/New/Project**" from the Capture menu, as shown in Figure 1.1. A new project dialog box will pop up and you will select "**Analog or Mixed A/D**." The "New Project" requires a project name and location. The location is the name of the subdirectory where the project file should be stored. ORCAD uses ".opj" as the file extension. Select "**OK**" in the "New Project" dialog box. A dialog box under "**Create PSPICE Project**" will pop up. Select "**Create a blank project**." The sequence of steps for starting ORCAD Schematic is shown in Box 1.1.

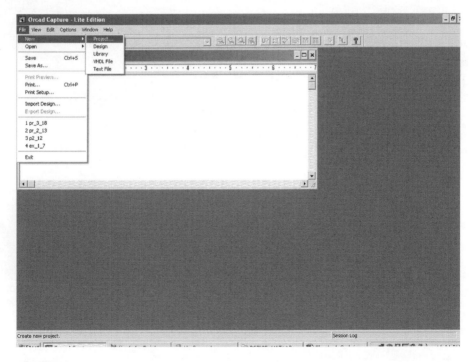

FIGURE 1.1
ORCAD capture opening screen.

> ### BOX 1.1　SEQUENCE OF STEPS FOR STARTING ORCAD SCHEMATIC
>
> - Open all Programs.
> - Go to ORCAD Family Release 9.2 program.
> - Click on Capture Lite Editor.
> - Open File/New/Project.
> - Select Analog or Mixed A/D.
> - Provide Name of the Project and location of the project file.
> - Click OK.
> - In the "Create PSPICE Project" (for new project) dialog box, select "Create a blank project."

1.2.2 Drawing a Circuit Using the ORCAD Schematic

The following three activities are needed to draw a circuit: (i) placing the circuit elements in the Capture Workspace, (ii) adjusting the circuit elements parameter values, and (iii) connecting the circuit elements by wires. The following example will illustrate the use of the ORCAD schematic capture.

Example 1.1: Schematic of a Simple Passive Circuit

Suppose we want to draw the circuit shown in Figure 1.2 and obtain the nodal voltages of the circuit. Proceed with the following steps:

(a) Placing Circuit Elements in the Capture Workspace

Select "**Place/Part**" from the ORCAD Capture menu. A "Place Part" dialog box will show up, click on "**Add Library.**" ORCAD Capture has several libraries. These include: analog.olb, breakout.olb, source.olb, and special. olb. Select the library, "analog.olb," and click on the "**Open**" button to have the elements in the library available to you for drawing. In addition, select the library, "source.olb," to have its elements available to you.

To obtain a resistor from the libraries, select "**ANALOG**" from the libraries list, and R from the "**Part List.**" Select "**OK**" to close the "Place Part" dialog box. The resistor is placed at the desired location by using the left mouse click. Table 1.1 shows some of the electrical parts and the corresponding libraries where they can be found.

SPICE requires a ground node for each circuit. The ground can be selected from "**Place/Ground**" from the ORCAD Capture menu.

(b) Adjusting Values of the Parameters of Circuit Elements

All the circuit elements have their default values. Left click on the part to select the element and then right click to perform specific functions such as edit properties, mirror vertically, and mirror horizontally. The suffix letters

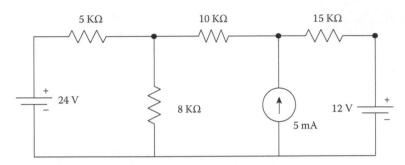

FIGURE 1.2
Electric circuit with active and passive elements.

TABLE 1.1

PSPICE Parts and Corresponding Libraries

Description	PSPICE Name	PSPICE Library
Resistor	R	ANALOG
Capacitor	C	ANALOG
Inductor	L	ANALOG
DC voltage source	VDC	SOURCE
DC current source	IDC	SOURCE

TABLE 1.2

Abbreviations of SPICE Scaling Factor

Suffix Letter	Metric Prefix	Multiplying Factor
T	Tera	10^{12}
G	Giga	10^{9}
Meg	Mega	10^{6}
K	Kilo	10^{3}
M	Milli	10^{-3}
U	Micro	10^{-6}
N	Nano	10^{-9}
P	Pico	10^{-12}
F	Femto	10^{-15}
Mil	Millimeter	$25.4 * 10^{-6}$

or multiplying factors may be used to adjust the values of parameters of the circuit elements. Table 1.2 shows the PSPICE scale factors and their abbreviations.

You can rotate the element by left clicking on it to select it, and then right clicking on it to rotate it. The elements R2 and I1 were rotated to obtain the figure shown in Figure 1.3.

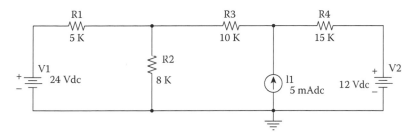

FIGURE 1.3
Circuit of Figure 1.2 as drawn in ORCAD capture.

BOX 1.2   STEPS FOR DRAWING CIRCUITS BY USING ORCAD SCHEMATIC

- Select "Place/Part" from ORCAD Capture menu.
- Click on "Add Library."
- Select library, "Analog" and click on "open" button.
- Select library, "source" and click on "open" button.
- Click on "Open" button.
- Specific parts are selected from the appropriate library and place.
- Select part (i.e., R, C, L) from the Analog Library.
- Select part (i.e., V, I) from the Source Library.
- Click "OK" to close "Place/Part" dialog box.
- To rotate an element, left click on it to select it and right click to rotate it.
- To adjust the element values, right click on the value, and left click to open the properties menu. Right click on the edit properties to change the element values, or double click on the value and change the value.
- To wire the circuit, select "Place/Wire," left press the mouse on one of the squares on the element and drag the mouse to the other terminal. Release the mouse by left pressing the mouse again. Continue this operation until all the elements are connected.
- Right click to bring up the "End Wire" menu and click on "end wire" when wiring is complete.
- Double click on "GND" and the "Property Editor" menu will show up. Under "NAME," change "GND" to 0, but do not change the entry in the source symbol.

(c) Wiring the Circuit

To connect the circuit elements, select "Parts/Wire." Left press the mouse on one of the squares on an element and drag the mouse to the other terminal and then release the mouse by left pressing the mouse again. Box 1.2 shows the steps for drawing circuits by using PSPICE schematics.

(d) Simulating the Circuit

"PSPICE/New Simulation Profile" from ORCAD Capture menu is selected to begin the simulation. A "Simulation Setting" dialog box will pop up. In the "New Simulation" dialog box, type in the name for the simulation and click on "create." Select "Bias Point" for the analysis type and under options, select "General Settings." To run the simulation, select "PSPICE /Run."

(e) Displaying Simulation Results

The results of simulations can be obtained in a text file or a plot. You can examine the output text file by clicking on the third button on the left vertical toolbar. On the PSPICE A/D, the simulation results can also be seen from "View/Output File." You can also see node voltages displayed directly on the schematic. To do this you exit the PSPICE A/D and return to the schematic. Click on the "V" symbol in the second row toolbar. The simulated circuit is shown in Figure 1.4.

1.3 DC Analysis

Under DC analysis, PSPICE can perform (i) DC nodal analysis and (ii) DC sweep. In nodal analysis, PSPICE determine the values of the node voltages and also the values of current in voltage sources. This type of analysis is also described as "Bias Point" analysis in PSPICE. The two types of DC analysis are described below in Sections 1.3.1 and 1.3.2. Some of the circuits for DC analysis might contain dependent sources. The list of PSPICE dependent sources and the library they can be found in is shown in Table 1.3.

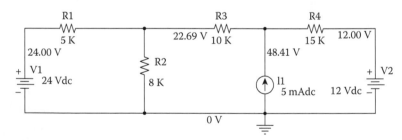

FIGURE 1.4
Simulation results of Figure 1.2 (nodal voltages are shown).

TABLE 1.3

PSPICE Dependent Sources

Description	PSPICE Name	Library
Voltage-controlled voltage source (VCVS)	E	ANALOG
Current-controlled current source (CCCS)	F	ANALOG
Voltage-controlled current source (VCCS)	G	ANALOG
Current-controlled voltage source (CCVS)	H	ANALOG

1.3.1 Bias Point Calculations

In Bias point calculations, the DC voltages at various nodes and current flowing through circuit elements are obtained. The following example illustrates the Bias point calculations.

Example 1.2: Bias Point Calculation of a Circuit with Dependent Source

Consider the circuit shown in Figure 1.5. The dependent source is voltage-controlled current source. Determine the nodal voltages.

Solution

The schematic drawing was done and the values of the circuit parameters were adjusted by using the steps in Box 1.2. The circuit elements were wired together. The gain of the voltage-controlled current source is changed in the Property editor, shown in Figure 1.6.

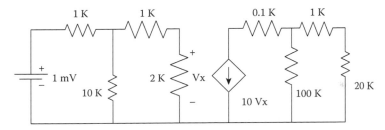

FIGURE 1.5
Circuit with voltage-controlled current source.

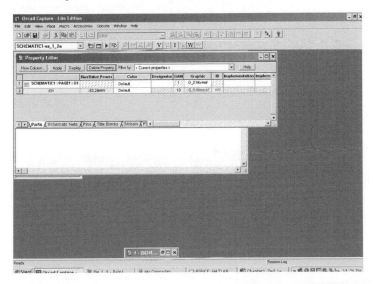

FIGURE 1.6
Setting the gain of the VCCS.

BOX 1.3   STEPS FOR PERFORMING ORCAD SCHEMATIC DC ANALYSIS

- Select "PSPICE/New Simulation Profile."
- Insert the name of the simulation in "New Simulation" menu.
- Click on "Create."
- In the "Simulation Settings," select "Bias Point" for analysis type, and choose "General Settings" under option.
- Select "OK" to close "Simulation Settings" dialog box.
- To run the DC analysis, choose "PSPICE/Run."
- View the simulation results either by choosing "View/Output" for the output text file or by going to ORCAD Capture and selecting "V" to view the nodal voltages.

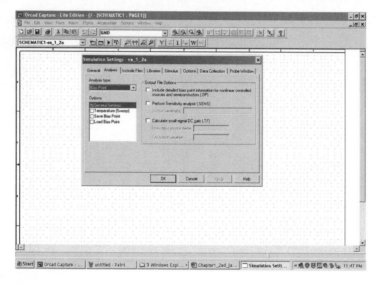

FIGURE 1.7
Simulation settings for bias point calculations.

To simulate the circuit for DC analysis, the sequence of actions shown in Box 1.3 is used to simulate the circuit. Figure 1.7 shows the simulation settings for the bias calculations. Figure 1.8 shows the voltages at various nodes of Figure 1.6 upon performing the DC analysis.

1.3.2 DC Sweep

In a DC sweep, one or more of the DC voltages are allowed to change and the voltages at nodes and currents through devices are monitored. The analysis type used is DC sweep. The following example illustrates the use of the DC sweep.

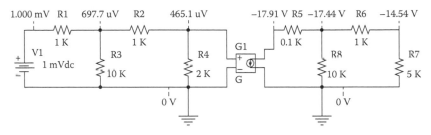

FIGURE 1.8
DC analysis results.

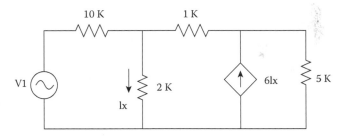

FIGURE 1.9
Circuit for DC sweep.

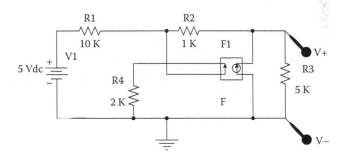

FIGURE 1.10
Schematic drawing of Figure 1.9.

Example 1.3: DC Sweep Analysis

For the circuit shown in Figure 1.9, plot the output voltages as the voltage V1 changes from 5 V to 10 V.

Solution

The schematic diagram of circuit was drawn by using the steps outlined in Box 1.2. You should note that the "edit properties" was used to change the scaling factor of the current controlled current source to 6. The schematic drawing is shown in Figure 1.10.

To perform the DC sweep simulations, the following steps are used. Select PSPICE/New Simulation Profile. In the New Simulation dialog box, specify the simulation name and select "create." In the "Simulation Settings" dialog, select "DC Sweep." For the sweep variable, use the voltage V1. Select the sweep type to be linear, the starting value of 5 V, end value of 10 V, and increments of 0.5 V. Figure 1.11 shows the simulation settings for the DC sweep simulation. The steps for performing the DC Sweep simulation are summarized in Box 1.4.

If the DC sweep simulation is successful, a PROBE (to be discussed in the Section 1.4) screen will open in the schematic window. To display the results of the simulation, indicate the variables to be displayed and the manner of displaying them. In this example, we want to display Vout versus Vin. In ORCAD schematic, put the voltage difference marker at the nodes of interest. If we were interested in current through an element, we place the current marker button on the element

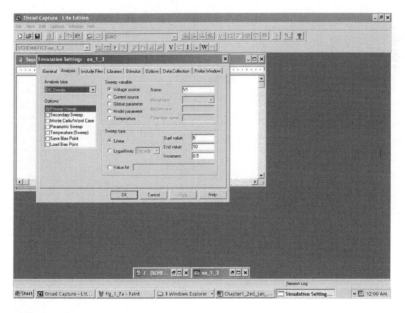

FIGURE 1.11
Simulation profile for the DC sweep simulation.

BOX 1.4   STEPS FOR PERFORMING ORCAD SCHEMATIC DC SWEEP ANALYSIS

- Select PSPICE/New Simulation Profile.
- In the "New Simulation" dialog box, specify Simulation name.
- Select "Create."
- In the "Simulation Settings" dialog box, select DC Sweep.
- Select the sweep variable.
- Choose sweep type, start value, end value, and increment.
- Select "OK" to close the "Simulation Settings" dialog box.
- To run DC analysis, choose "PSPICE/Run."

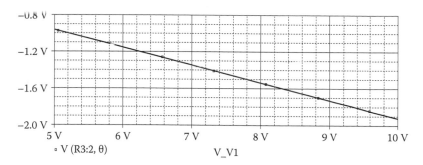

FIGURE 1.12
Output voltage of Figure 1.9.

of interest. Another way to display the variables would be to select "Trace/Add Trace" to pop up the "Add Traces" dialog box. You can then select the voltages or currents of interest to you. The output voltage with respect to the input voltage is shown in Figure 1.12.

1.4 PROBE

PROBE is a PSPICE interactive graphics processor that allows the user to display SPICE simulation results in graphical format on a computer monitor. PROBE has facilities that allow the user to access any point on a displayed graph and obtain its numerical values. In addition, PROBE has many built-in functions that enable a user to compute and display mathematical expression that models aspects of circuit behavior.

PROBE has several commands available for file accessing, plotting, editing, viewing, and adding or removing trace. PSPICE has several functions that PROBE can use to determine various characteristics of a circuit. Table 1.4 shows the valid functions for PROBE expression.

1.5 Transient Analysis

Transient analysis is used to obtain the response of a circuit to a time varying input. Circuits with energy storage elements are of special interest with respect to transient analysis. The capacitors or inductors should have initial conditions for transient analysis. PSPICE-supplied sources for transient analysis are shown in Table 1.5. The details of the sources will be described in Chapter 2.

The following example uses rectangular pulse as the input to a RC circuit.

TABLE 1.4

Valid Functions for PROBE Expression

Function	Meaning	Example
+	Addition of current or voltage	V(3) + V(2,1) + V(8)
−	Subtraction of current or voltage	I(VS4) − I(VM3)
*	Multiplication of current or voltage	V(11)* V(12)
/	Division of current or voltage	V(6)/V(7)
ABS(X)	$|X|$, Absolute value of X	ABS(V(9))
SGN(X)	+ 1 if X > 0; 0 if X = 0; −1 if X < 0	SGN(V(4))
SQRT(X)	$X^{1/2}$, square root of X	SQRT(I(VM1))
EXP(X)	e^X	EXP(V(5,4))
LOG(X)	Ln(X), log base e of X	LOG(V(9))
LOG10(X)	$Log_{10}(X)$, log base 10 of X	LOG10(V(10))
DB(X)	$20 * log_{10}(X)$, magnitude in decibels	DB(V(6))
PWR(X,Y)	$|X|^Y$, X to the power Y	PWR(V(2),3)
SIN(X)	sin(X), X in radians	SIN(6.28 * V(2))
COS(X)	cos(X), X in radians	COS(6.28 * V(3))
TAN(X)	tan(X), X in radians	TAN(6.28 * V(4))
ARCTAN(X)	$tan^{-1}(X)$, X in radians	ARCTAN(6.28 * V(2))
ATAN	$tan^{-1}(X)$, results in radians	ATAN(V(9)/V(4))
d(X)	Derivative of X with respect to X-axis variable	D(V(12))
S(X)	Integral of X over the X-axis variable	S(V(15))
AVG(X)	Running average of X over the range of X-axis variable	AVG(V5,3))
*AVGX(XO,XF)	*Running average of X from X-axis value, XO, to the x-axis value, XF.	AVG V(5,4)(2e-3,20e-3)
RMS(x)	Running RMS average of X over the range of the X-axis variable	RMS(VS2)
MIN(X)	Minimum of real part of X	MIN(VM3)
MAX(X)	Maximum of real part of X	MAX(VM3)
M(X)	Magnitude of X	M(V(5))
P(X)	Phase of X, result in degrees	P(V(4))
R(X)	Real part of X	R(V(3))
IMG(X)	Imaginary part of X	IMG(V(6))
G(X)	Group delay of X, results in seconds	G(V(7))

TABLE 1.5

SPICE-Supplied Sources for Transient Analysis

Name	Application
PULSE	For periodic pulse waveforms
EXP	For exponential waveforms
PWL	For piecewise-linear functions
SIN	For sinusoidal waveforms
SFFM	For frequency-modulated waveforms

Example 1.4: Transient Analysis of RC Circuit

For the RC circuit shown in Figure 1.13, the input signal is a pulse. Plot the voltage $v_o(t)$ with respect to time.

The pulse is described by VPULSE. The parameters for the VPULSE are described by V1, V2, TD, TR, TF, PW, and PER,

where

V1 is the initial value of pulse. There is no default value for V1.

V2 is the final voltage of the pulse. There is no default value for V2.

TD is delay time, its default value is zero.

TR is the rise time. Its default value is the printing or plotting increment.

TF is the fall time. Its default is also TSTEP.

PW is the pulse width. The default value of PW is TSTOP, the final time of the transient analysis.

PER is the period. Its default is also TSTOP. The period does not include the initial delay, TD.

In this example, $V1 = 0$, $V2 = 5$ V, $TD = 0$, $TR = 10^{-6}$ s, $TF = 10^{-6}$ s, $PW = 5$ ms, and $PER = 20$ ms.

Solution

Using the steps outlined in Box 1.2, the circuit is drawn using ORCAD capture. For the voltage source, use the ORCAD property editor to edit the list of parameters for the "PULSE" voltage source. The schematic is shown in Figure 1.14.

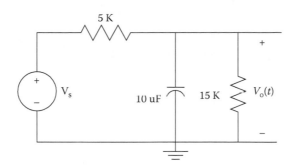

FIGURE 1.13
RC circuit with an input pulse voltage.

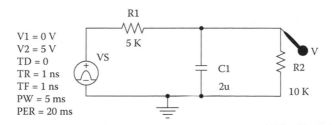

FIGURE 1.14
ORCAD capture drawing of Figure 1.13.

To simulate, select "PSPICE/New Simulation Profile," specify the simulation name and select "create." In the simulation settings dialog choose "Time Domain (Transient)." The simulation starts at time equal to zero and ends at run to time. Select "skip the initial transient bias point calculations (skip BP)." Figure 1.15 shows the simulation profile for the transient analysis. Box 1.5 shows the steps for performing the transient analysis.

If the simulation is successful, the PROBE window will open up automatically. Select "trace/Add Trace" and choose the output. The output voltage is shown in Figure 1.16.

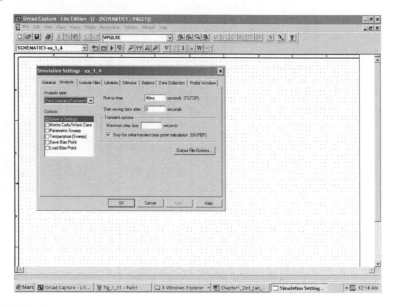

FIGURE 1.15
Simulation profile for transient analysis.

BOX 1.5   STEPS FOR PERFORMING ORCAD SCHEMATIC TRANSIENT ANALYSIS

- Select PSPICE/New Simulation Profile.
- In the New Simulation dialog box, specify Simulation name.
- Select "Create."
- In the "Simulation Settings" dialog box select "Time Domain (Transient)" in the analysis type and select "General Settings" under options.
- Choose the start and end time of simulation.
- Skip the initial transient bias calculations "SKIPBP," if the initial bias points are known.
- Select "OK."
- Select "PSPICE/Run."

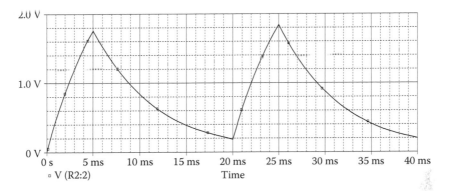

FIGURE 1.16
Output voltage of Figure 1.14.

TABLE 1.6

PSPICE Switches

Description	PSPICE Name	PSPICE Library
Voltage-controlled switch	S	ANALOG
Switch is normally opened. It will close at time TCLOSE	Sw_tClose	EVAL
Switch is normally closed. It will open at time TOPEN	Sw-tOpen	EVAL

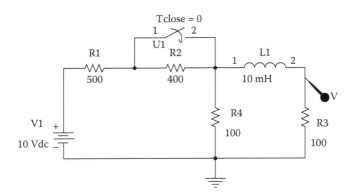

FIGURE 1.17
RL circuit with a switch.

Transient analysis sometimes involves the use of switches. PSPICE has three switches for time-domain analysis. They are shown in Table 1.6.

The following example illustrates the use of one of the switches.

Example 1.5: RL Circuit with a Switch

Figure 1.17 shows an RL circuit. The switch closes at $t = 0$ seconds. Find the voltage across the resistance R3.

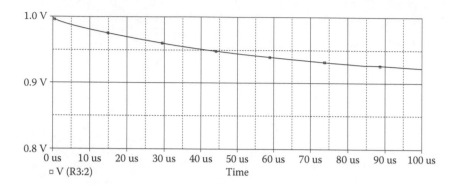

FIGURE 1.18
Voltage across the resistor R3.

Solution

Use ORCAD Capture to place the parts, adjust the parameter values and wire the parts together (see Box 1.2). To simulate the circuit, select the steps in Box 1.5 to perform the transient analysis. However, in the "Simulation Settings" dialog box, do not skip the bias point calculation. You can use run time to be 100 μs. The simulation results are shown in Figure 1.18.

1.6 AC Analysis

For time-invariant circuit, excited by a sinusoidal source, AC analysis can be used to obtain voltages and currents in the circuit. For circuits with more than one input, the superposition theorem may be used to obtain the response. In AC analysis, the voltages and currents are transformed into the frequency domain. The calculations are performed using phasors (magnitudes and phases of the voltages and currents).

PSPICE has voltage source (VAC) and current source (IAC) for AC analysis. The voltage and current sources are available in the SOURCE library. The magnitude and phase of the AC sources are set using the property editor. However, the frequency of the AC sources is specified using the "Simulation Settings" dialog box. The IPRINT and VPRINT are used to print current and voltage, respectively. IPRINT is connected in series with the circuit to measure current. However, VPRINT is connected in parallel across an element whose voltage needs to be obtained. The passive sign convention is used to describe the polarity of the voltage and current printers (VPRINT and IPRINT). The minus sign on the printer VPRINT indicates the negative polarity of the voltage. For the IPRINT, the minus sign indicates the node the current leaves the printer. The following example shows how to perform AC analysis on a simple RLC circuit.

Example 1.6: AC Analysis of a RLC Circuit

Figure 1.19 shows an RLC circuit. If $v(t) = 18\cos(200\pi t + 60°)$, find $i_c(t)$ and $v_{R1}(t)$.

Solution

Using the information in Box 1.2, the circuit is drawn. The ORCAD schematic drawing is shown in Figure 1.20. The IPRINT and VPRINT have been inserted to obtain $i_c(t)$ and $v_{R1}(t)$, respectively.

The property editor can be used to change the property of AC voltage (VAC) to be displayed by using the display button. To change the properties of each printer, left click on the printer to select it, and then right click on the printer to bring up a menu. Select "Edit Properties" and set the AC fields, IMAG, MAG, and PHASE to "y." If for some reason, one of the fields, REAL, IMAG, MAG, and PHASE, cannot be found in the property editor, "Add New Column" dialog box will pop up and the field can be added.

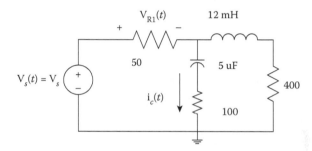

FIGURE 1.19
RLC circuit.

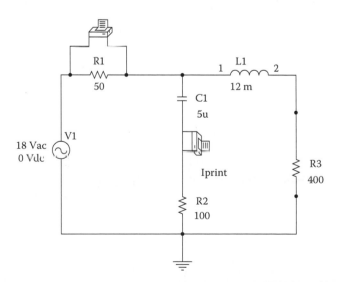

FIGURE 1.20
ORCAD schematic capture diagram of Figure 1.19.

To simulate the circuit, select "PSPICE/New Simulation Profile." A dialog box will pop up. Under "New Simulation" dialog box, specify the simulation name and click "Create." Under "Simulation Settings" select "AC Sweep/Noise" as the analysis type. Since the circuit needs to be simulated at one frequency, make the start frequency the same as the end frequency. It is 1000 Hz. The start and end frequencies should be in Hertz. Run the simulation using the "PSPICE/Run." The simulation profile for the AC analysis is shown in Figure 1.21. Box 1.6 summarizes the steps needed to perform AC Analysis in ORCAD Capture.

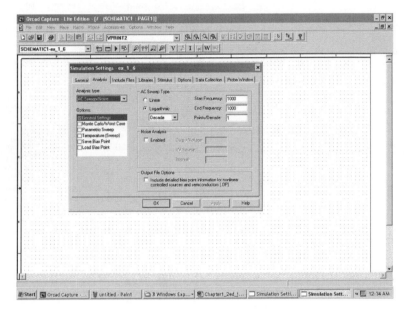

FIGURE 1.21
Simulation profile for AC analysis.

BOX 1.6   STEPS FOR PERFORMING ORCAD SCHEMATIC AC ANALYSIS

- Select PSPICE/New Simulation Profile.
- In the New Simulation dialog box, specify Simulation name.
- Click "Create."
- In the "Simulation Settings" dialog box, select "AC Sweep/Noise" as the analysis type.
- Enter the start and end frequencies.
- Select "OK" to close the "Simulation Settings" dialog box.
- Run the simulation using "PSPICE/Run."
- Select "View/Output File" to view the results of the AC Analysis.

The results of the simulation can be obtained from the output file. It is shown in Table 1.7.

From the PSPICE results, we get

$$i_c(t) = 0.1086\cos(2000\pi t + 73.24°) \text{ A, and}$$

$$v_{R1}(t) = 6.69\cos(2000\pi t + 67.54°) \text{ V.}$$

Frequency response of circuits can be obtained by performing the AC analysis. The AC sweep is used to simulate circuits over a range of frequencies. The start and end frequencies determine the range of the AC sweep. The following example illustrates the use of the AC sweep to determine the frequency response of a filter.

Example 1.7: RLC Filter

The circuit shown in Figure 1.22 is a passive filter. Plot the magnitude of the output voltage $v_o(t)$.

TABLE 1.7

AC Analysis Results from PSPICE Simulations

**** AC ANALYSIS		TEMPERATURE = 27.000 DEG C	
FREQ	IM(V_PRINT1)	I P(V_PRINT1)	IR(V_PRINT1)
1.000E + 03	1.086E-01	7.324E + 01	3.132E-02
FREQ	VM(N00491,N00525)	VP(N00491,N00525)	VR(N00491,N00525)
1.000E + 03	6.695E + 00	6.754E + 01	2.558E + 00

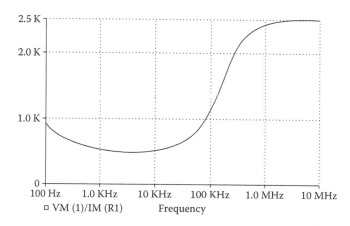

□ VM (1)/IM (R1)   Frequency

FIGURE 1.22
A passive filter.

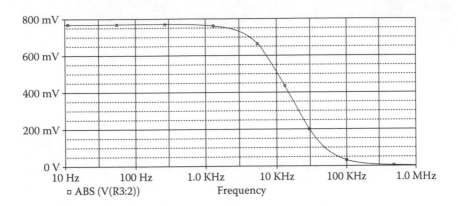

FIGURE 1.23
Magnitude response of output voltage $v_o(t)$.

Using the information in Box 1.2, Figure 1.22 is drawn using ORCAD Capture. In addition, by using the steps in Box 1.6, you can perform the AC analysis. However, the start frequency and end frequency are selected to allow the sweep over a range of frequencies. The frequency response is shown in Figure 1.23.

Problems

1.1   Find the nodal voltages of the circuit shown in Figure P1.1.

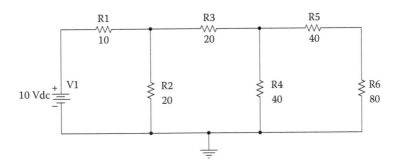

FIGURE P1.1
Resistive circuit.

1.2   Use PSPICE to determine the currents flowing through the 20 V source and the resistor R6.

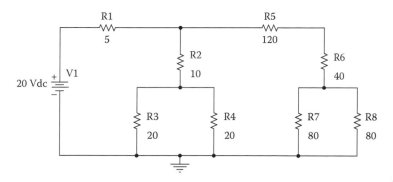

FIGURE P1.2
Resistive circuit.

1.3 Find the current I_s.

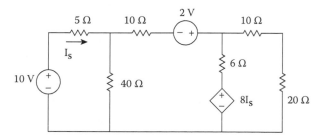

FIGURE P1.3
Passive circuit with dependent sources.

1.4 For the resistive capacitive circuit shown in Figure P1.4, the source is a pulse with amplitude of 5 V and pulse duration of 10 ms. Determine the voltage across the capacitor.

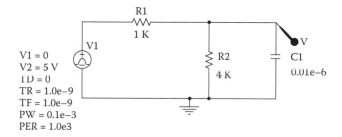

FIGURE P1.4
Resistive capacitive circuit.

1.5   Find the current flowing through the inductor if the source is a periodic pulse signal with pulse duration of 0.4 ms, period of 1.0 ms, and pulse amplitude of 10 mA.

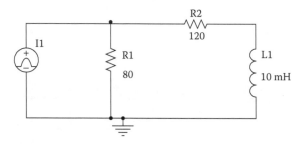

FIGURE P1.5
Resistive inductive circuit.

1.6   For the circuit shown in Figure P1.6, the current flowing through the inductor is zero at time $t < 0$. At $t = 0$, the switch moved from position a to b, where it remained for 10 ms. After the 10 ms delay, the switch moved from position b to position c, where it remained indefinitely. Sketch the current flowing through the inductor versus time.

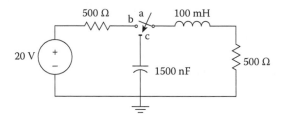

FIGURE P1.6
RLC circuit.

1.7   For the voltage shown in Figure P1.7, plot the magnitude response of the voltage across the inductor.

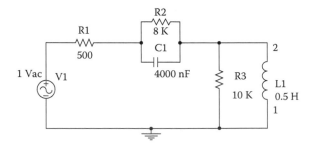

FIGURE P1.7
RLC circuit.

1.8   The input voltage is given as $v1(t) = 50\cos(1000\pi t)$, determine the voltage $v_o(t)$.

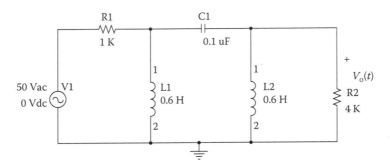

FIGURE P1.8
RLC circuit for AC analysis.

Bibliography

1. Nilsson, James W., and Susan A. Riedel. *Introduction to PSPICE manual Using ORCAD Release 9.2 to Accompany Electric Circuits.* Upper Saddle River, NJ: Pearson/Prentice Hall, 2005.
2. OrCAD Family Release 9.2. San Jose, CA: Cadence Design Systems, 1986–1999.
3. Rashid, Mohammad H. *Introduction to PSPICE Using OrCAD for Circuits and Electronics.* Upper Saddle River, NJ: Pearson/Prentice Hall, 2004.
4. Soda, Kenneth J. "Flattening the Learning Curve for ORCAD-CADENCE PSPICE," *Computers in Education Journal,* Vol. XIV (April–June 2004): 24–36.
5. Svoboda, James A. *PSPICE for Linear Circuits.* 2nd ed. New York: John Wiley & Sons, Inc., 2007.
6. Tobin, Paul. "The Role of PSPICE in the Engineering Teaching Environment." Proceedings of International Conference on Engineering Education, Coimbra, Portugal, September 3–7, 2007.
7. Tobin, Paul. *PSPICE for Circuit Theory and Electronic Devices.* San Rafael, CA: Morgan & Claypool Publishers, 2007.
8. Tront, Joseph G. *PSPICE for Basic Circuit Analysis.* New York: McGraw-Hill, 2004.
9. Vladimirescu, Andrei. *The Spice Book.* New York: John Wiley and Sons, Inc., 1994.
10. Wyatt, Michael A. "Model Ferrite Beads in SPICE." In *Electronic Design,* October 15, 1992.
11. Yang, Won Y., and Seung C. Lee. *Circuit Systems with MATLAB® and PSPICE.* New York: John Wiley & Sons, 2007.

2

PSPICE Fundamentals

2.1 Introduction

In Chapter 1, a schematic input file was created from circuit schematic. From the schematics file, a netlist was created automatically. Using PSPICE A/D, all the voltages in the circuit are obtained. The simulation results were processed using PROBE. Another way of performing PSPICE simulations is to start with a circuit file. In circuit files, the user assigns node numbers, specifies the elements connected to the various nodes, the type of simulation to be performed, and the output to be printed or plotted.

For many circuit simulations, working from the circuit schematic is much more convenient than performing the simulations using circuit files. However, for circuits that are very large and require advanced commands, the generation of circuit files is often necessary. This chapter describes PSPICE simulations using circuit files.

A general SPICE circuit file program consists of the following components:

- Title
- Element statements
- Control statements
- End statements

The following two sections will discuss the element and control statements.

2.1.1 Element Statements

The element statements specify the elements in the circuit. The element statement contains the (a) element name, (b) the circuit nodes to which each element is connected, and (c) the values of the parameters that electrically characterize the element.

The element name must begin with a letter of the alphabet that is unique to a circuit element, source, or subcircuit. Table 2.1 shows the beginning alphabet of an element name and the corresponding element.

TABLE 2.1

Element Name and Corresponding Element

First Letter of Element Name	Circuit Element, Sources, and Subcircuit
B	GaAs MES field-effect transistor
C	Capacitor
D	Diode
E	Voltage-controlled voltage source
F	Current-controlled current source
G	Voltage-controlled current source
H	Current-controlled voltage source
I	Independent current source
J	Junction field-effect transistor
K	Mutual inductors (transformers)
L	Inductor
M	MOS field-effect transistor
Q	Bipolar junction transfer
R	Resistor
S	Voltage-controlled switch
T	Transmission line
V	Independent voltage source
X	Subcircuit

Circuit nodes are positive integers. The nodes in the circuit need not be numbered sequentially. Node 0 is predefined for ground of a circuit. To prevent error messages, all nodes must be connected to at least two elements.

Element values can be an integer, floating pointer number, integer floating point followed by an exponent, or floating point or integer followed by scaling factors shown in Table 1.2. Any character after the scaling factor abbreviation is ignored in SPICE. For example, a 5000 Ohm resistor can be written as 5000 or 5000.00 Ohm, 5 K, 5 E3, 5 Kohm, or 5 KR.

The element statements of some common elements such as a resistor, inductor, capacitor, independent voltage source, and independent current source will now be described.

Resistors

The general format for describing resistors is

Rname N + N– value [TC = TC1,TC2]

where

The name must start with the letter **R;**

N + and **N–** are the positive and negative nodes of the resistor. Conventional current flows from the positive node N + through the resistor to the negative node, N–;

Value specifies the value of the resistor. The latter may be positive or negative, but not zero; and

TC1 and **TC2** are the temperature coefficients. The default values are zero. If they are nonzero, then the resistance is given by the formula:

$$\text{Resistor value} = value\ [1 + TC1(T - T_{nom}) + TC2(T - T_{nom})^2] \qquad (2.1)$$

where
  TC1 is linear temperature coefficient;
  TC2 is quadrature temperature coefficient; and
  T_{nom} is the nominal temperature set using TNOM option. Its default value is 27°C.

Inductors
The general format for describing linear inductors is:

Lname N + N− value [IC = initial_current]

where
  The inductor name must start with the letter **L;**
  N + and **N−** are positive and negative nodes of the inductor, respectively. Conventional current flows from the positive node to the negative node;
  Value specifies the values of the inductance; and
  The initial condition for transient analysis is assigned using **IC** = initial_current to specify the initial current.

Capacitors
The general format for describing linear capacitors is:

Cname N + N− value [IC = initial_voltage]

where
  The capacitor name must start with the letter **C;**
  N + and **N−** are the positive and negative nodes of the capacitor, respectively;
  Value indicates the value of the capacitance; and
  The initial condition for transient analysis is assigned using **IC** = initial_voltage on the capacitor.

Independent Voltage Source
The general format for describing independent voltage source is:

Vname N + N− [DC value] [AC magnitude phase]

[PULSE V$_1$ V$_2$ td tr tf pw per]

or [SIN V$_O$ V$_a$ freq td df phase]

or [EXP V$_1$ V$_2$ td$_1$ t$_1$ td$_2$ t$_2$]

$$\text{or } [\text{PWL t1 } V_1 t_2 V_2 \dots t_n, V_n]$$

$$\text{or } [\text{SFFM } V_O V_a \text{ freq md fs}]$$

where
The voltage source must start with letter **V**;
$N+$ and $N-$ are the positive and negative nodes of the source, respectively; and
Sources can be assigned values for DC analysis **[DC value]**; AC analysis **[AC Magnitude phase]**, and transient analysis. Only one of the transient response source options (**PULSE, SIN, EXP, PWL, SFFM**) can be selected for each source. The AC phase angle is in degrees. The transient signal generators, PULSE, SIN, EXP, PWL, SFFM will be discussed in Section 2.5.

Independent Current Source

The general format for describing independent current source is:

Iname $N+N-$ [DC value] [AC magnitude phase]

$$[\text{PULSE } V_1 V_2 \text{ td tr tf pw per}]$$

$$\text{or } [\text{SIN } V_O V_a \text{ freq td df phase}]$$

$$\text{or } [\text{EXP } V_1 V_2 \text{ td}_1 t_1 \text{ td}_2 t_2]$$

$$\text{or } [\text{PWL } t_1 V_1 t_2 V_2 \dots t_n, V_n]$$

$$\text{or } [\text{SFFM } V_O V_a \text{ freq md fs}]$$

where
The current source must start with letter **I**;
$N+$ and $N-$ are the positive and negative nodes of the source, respectively. Current flows from a positive node to the negative node; and
Independent current sources can be assigned values for DC analysis **[DC value]**; AC analysis **[AC Magnitude phase]**, and transient analysis. Only one of the transient response source options (**PULSE, SIN, EXP, PWL, SFFM**) can be selected for each source. The AC phase angle is in degrees.

2.1.2 Control Statements

Circuit Title

The circuit title must be the first statement in the SPICE circuit file program or circuit netlist. If this is not done, the program will assume the first statement is the circuit title. The circuit title is used to label the output when the analysis is complete. If additional comments are needed in the circuit description, the comment statement can be used.

Comments (*)

An asterisk, *, in the first column of a line indicates a comment line. An example of a comment is as follows:

```
*

* Comment line begins with asterisks in SPICE.

*
```

Operating Point (.OP)

The operating point of devices and elements in a circuit can be printed out using the .OP command. The general format for using the .OP control statement is:

$$.OP$$

PSPICE always calculates the operating point of devices in a circuit. With .OP control statement, the following values are printed:

a. voltages at each node of a circuit;

b. currents and power dissipation of all voltage sources in a circuit; and

c. transistor diode parameters, if the previously mentioned devices are present in a circuit.

Other control statements, such as .DC (DC analysis), .TRAN (Transient analysis), and .AC (AC analysis) will be discussed in the following sections. Additional control statements will be covered in Chapter 2. The following example illustrates the .OP control statement and the element statements.

Example 2.1: Resistive Circuit with Multiple Sources

Figure 2.1 shows a resistive circuit with multiple sources. VS = 10 V, R1 = 500 Ω, R2 = 1 KΩ, R3 = 2 KΩ, R4 = 1 KΩ, R5 = 3 KΩ, R6 = 5 KΩ, and I1 = 5 mA. Find the nodal voltages.

Solution

PSPICE Circuit File Program

```
Resistive Circuit with Multiple Sources
VS 1 0 DC        10V
R1 1 2 500
R2 2 3 1000
R3 3 0 2000
R4 2 4 1000
R5 4 5 3000
R6 5 0 5000
I1 3 5 DC        5mA
.OP
.END
```

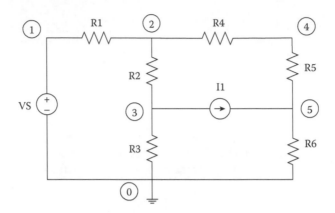

FIGURE 2.1
Resistive circuit with multiple sources.

The nodal voltages obtained from PSPICE output file are:

Node	Voltage, V
1	10.0000
2	7.9545
3	1.9697
4	9.8485
5	15.5300

2.2 DC Analysis

The .DC control statement specifies the values that will be used for DC sweep or DC analysis. The general format for the .DC statement is:

**.DC SOURCE_NAME START-VALUE STOP_VALUE
INCREMENT_VALUE**

where

SOURCE_NAME is the name of an independent voltage or
current source

START_VALUE, STOP_VALUE, and INCREMENT_VALUE represent the
starting, ending, and increment values of the source, respectively.
For example,

.DC Vsource 0.5   5   .1

causes a DC source named Vsource to be swept, starting at a value of 0.5 V and stopping at 5 V, with an incremental step of 0.1 V. PSPICE analyzes the circuit at each value of Vsource.

To perform a DC analysis with DC source of one specified value, .DC statement can be used but start_value and stop_value are made equal to the specified value of the DC source. For example,

$$.DC \quad VCC \quad 5 \quad 5 \quad 1$$

causes a DC source named VCC, with a constant voltage of 5 V, to be used for DC analysis.

The .DC control statement can be used to sweep a second independent voltage source over a specified range of values. The general format for double sweep is

.DC S1 S1_start S1_stop s1_incr S2 S2_start   S2_stop S2_incr

where
   S1 is the name of first source. The source is swept from S1_start and stops
       at S1_stop with increments of S1_incr; and
   S2 is the name of the second source. It is swept from S2_start and sweep
       stops at S2_stop with increments of S2_incr.

The first sweep source S1 will be the "inner" loop implying that the entire first sweep will be performed for each value of the second sweep. The two-source sweep is useful in generating current versus voltage characteristics of semiconductor devices. For example,

.DC VCE 0 10V .2V IB 0mA 1mA .2mA

supplies two sources VCE and IB for sweep. VCE will vary from 0 to 10 V with .2 V increments, while IB will sweep from 0 mA to 1 mA with 0.2 mA increments. For each value of IB, VCE is a sweep from 0 V to 10 V. An example that involves two sweeps can be found in Section 8.1. The following example shows a single voltage source sweep.

Example 2.2: Bridge Circuit: Calculation of Bridge Current and DC Sweep

For the bridge circuit shown in Figure 2.2, R1 = 100 Ω, R2 = 100 Ω, R3 = 100 Ω, R4 = 400 Ω, R5 = 300 Ω, and R6 = 50 Ω. If the source voltage VS is swept from 0 V to 10 V in increments of 2 V, find the current IB.

Solution
Figure 2.2 has been redrawn with node numbers and element names. The redrawn circuit is shown in Figure 2.3.

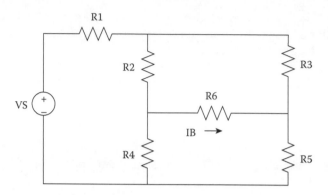

FIGURE 2.2
Bridge circuit.

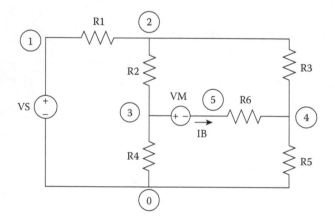

FIGURE 2.3
Figure 2.2 with node numbers and element names.

PSPICE Circuit File Program

```
Bridge Circuit
*
VS 1 0 DC 10V
VM 3 5 DC 0; current monitor
R1 1 2 100
R2 2 3 100
R3 2 4 100
R4 3 0 400
R5 4 0 300
R6 4 5 50
.DC VS 0 10 2
.PRINT DC I(VM)
.END
```

The relevant output from the program is:

Voltage Source VS, V	Current IB, A
0.000E+00	0.000E+00
2.000E+00	3.361E−04
4.000E+00	6.723E−04
6.000E+00	1.008E−03
8.000E+00	1.345E−03
1.000E+01	1.681E−03

2.3 Transient Analysis

The .TRAN control statement is used to perform transient analysis on a circuit. The general format of the .TRAN statement is:

.TRAN TSTEP TSTOP < TSTART > < TMAX > < UIC >

where

The terms inside the angle brackets are optional;

TSTEP is the printing or plotting increment;

TSTOP is the final time of the transient analysis;

TSTART is the starting time for printing out the results of the analysis. If it is omitted, it is assumed to be zero. The transient analyses always start at time zero. If TSTART is nonzero, the transient analysis computations are done from time zero to TSTART, but the results are not written to output file;

TMAX is the maximum step size that PSPICE uses for the purposes of computation. If TMAX is omitted, the default is the smallest value of either TSTEP or (TSTOP − TSTART)/50. TMAX is useful if you want the computational interval to be smaller than the TSTEP, the printing or plotting interval; and

UIC (Use Initial Conditions) is used to specify the initial conditions of capacitors and inductors. The initial conditions are specified in the element statement by adding term IC = value, for capacitors and inductors.

Before doing the transient analysis example, let us discuss the sources that can be used for transient analysis.

2.3.1 Transient Analysis Sources

There are five SPICE-supplied sources that can be used for transient analysis. They are:

PULSE < parameters > for periodic pulse waveform;

EXP < parameters > for exponential waveform;

PWL < parameters > for piece-wise linear waveform;

SIN < parameters > for a sine wave; and

SFFM < parameters > for frequency-modulated waveform.

The format for specifying the above sources for transient analysis are described below:

Pulse Waveform

The PULSE waveform is shown in Figure 2.4a. The general format of the pulse waveform is:

PULSE(V1 V2 td tr tf pw per)

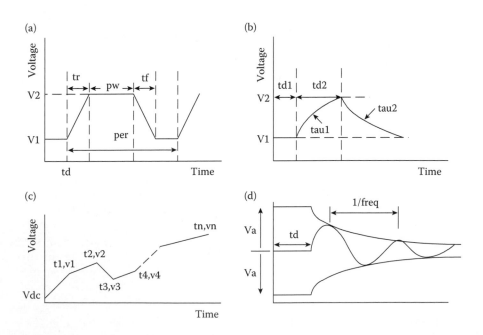

FIGURE 2.4

Transient analysis waveforms: (a) pulse waveform, (b) exponential waveform, (c) piece-wise linear waveform, and (d) exponentially damped sinusoid.

where

V1 is the initial value of pulse. There is no default value for V1;

V2 is the final voltage of the pulse. There is no default value for V2;

td is delay time. Its default value is zero;

tr is the rise time. Its default value is the printing or plotting increment;

tf is the fall time. Its default is also TSTEP;

pw is the pulse width. The default value of PW is TSTOP, the final time of the transient analysis; and

per is the period. Its default is also TSTOP. The period does not include the initial delay, td.

An example of using the PULSE statement is:

VPULSE 1 0 PULSE(0V 10V 10ns 20ns 50ns 1us 3us).

The above statement means a signal name VPULSE is connected to nodes 1 and 0. The pulse waveform starts at 0 V and stays there for 10 ns. The voltage increases linearly from 0 V to 10 V during the next 20 ns. The voltage stays at 10 V for 1 μs. Then, it decreases linearly from 10 V to 0 V during the next 50 ns. The cycle is repeated every 3 μs.

Exponential Waveform

An exponential waveform is shown in Figure 2.4b. The general format of the exponential waveform is:

EXP(V1 V2 td1 tau1 td2 tau2)

where

V1 is the initial voltage in volts. V1 must be specified since it does not have a default value;

V2 is the peak voltage in volts. It must also be specified;

td1 is the rise delay time in seconds. Its default value is zero;

tau1 is the rise time constant in seconds. Its default value is TSTEP, the printing or plotting increment in .TRAN statement;

td2 is the fall delay time in seconds. Its default is (td1 + TSTEP); and

tau2 is the fall time constant in seconds. The default of tau2 is TSTEP.

An example of using the EXP statement is:

VEXP 2 1 EXP(–1V 5V 1us 10us 30us 15us).

The above statement means the voltage VEXP, connected to nodes 2 and 1 is an exponential waveform. The waveform is –1 V for the first 1 μs. The voltage increases exponentially from –1 V to 5 V with a time constant of 10 μs. The voltage increase lasts for 30 μs. Then the voltage decays from 5 V to –1 V with a time constant of 15 us (see Figure 2.4).

Piecewise Linear Waveform

Piecewise linear function is constructed using straight lines between points. The piecewise linear waveform is shown in Figure 2.4c. The general format of the piecewise linear waveform is:

PWL (T1 V1 t2 V2 … Tn Vn)

where
   Each pair of time–voltage values, Tm, Vm, (where m = 1, 2, … n) specifies that the value of the source is Vm volts at time Tm seconds. The times are specified in increasing order. That is $t_1 \prec t_2 \prec t_3 \ \ldots \prec t_n$.

   An example of using the PWL statement is:

VPWL   1 0 PWL (0   0   1   2   4   2   5   3   7   3   8   2   11   2   12   0)

The above statement means the voltage VPWL connected to the nodes 1 and 0 is a piecewise linear function constructed from the following time–voltage values: (0, 0), (1, 2), (4, 2), (5, 3), (7, 3), (8, 2), (11, 2), and (12, 0).

Damped Sinusoidal Waveform

Sinusoidal source is generated using SIN. The exponentially damped sine wave is shown in Figure 2.4d. The general format of the sinusoid source is:

SIN (Vo Va freq td df phase)

where
   Vo is the offset voltage. It has no default value. It must be specified;
   Va is the peak amplitude. There is no default value. It must be specified;
   freq is the frequency. Its default value is 1/TSTOP. Where TSTOP is the
       final value of the transient analysis of the .TRAN statement;
   td is the delay time. Its default value is zero;
   df is the damping coefficient. Its default value is zero; and
   phase is the phase. Its default value is zero.

   The SIN function with its parameters can be used to generate the exponentially damped sine wave described by the equation.

$$v(t) = Vo + Va^*\sin[2\pi(\text{freq}(t - td)) - (\text{phase} / 360)]e^{-(t-td)df} \qquad (2.2)$$

An example of using the SIN statement is:

VSIN   2   1   SIN(0 10 10 K).

   The transient sinusoidal wave VSIN is generated with zero offset voltage, amplitude of 10 V and frequency of 10,000 Hz.
   It should be noted that the SIN waveform is for transient analysis only. It is deactivated during AC analysis.

Frequency-Modulated Sinusoidal Function

A single frequency-modulated signal is generated using SFFM function. The general format of the SFFM source is:

SFFM(Vo   Va   fc   mdi   fs)

where
   Vo is the offset voltage. It has no default and it must be specified;
   Va is the amplitude, it must also be specified;
   fc is the carrier frequency in Hertz. Its default is 1/TSTOP;
   mdi is the modulation index. Its default is zero; and
   fs is the signal frequency in Hertz.

The SFFM is described by the expression:

$$v(t) = Vo + Va*[\sin(2\pi*fc*t + mdi*\sin(2\pi*fs*t))] \tag{2.3}$$

An example of using SFFM function is:

VINPUT   4   0   SFFM(0   5   6Meg   8   20K)

The above source produces 6 MHz sinusoid with amplitude of 5 V modu-lated at 20 KHZ with a modulation index of 8. The following illustrates the use of transient response source being employed for transient analysis.

Example 2.3: Transient Response of a Series RLC Circuit

For the RLC circuit shown in Figure 2.5, L = 2 H, C = 1.5 μF and R = 1000 Ω. Find the voltage across the resistor if the input is a pulse waveform with pulse duration of 1 ms, and pulse amplitude of 5 V. Assume no initial conditions.

Solution

Figure 2.6 is Figure 2.5 with node numbers and element names.

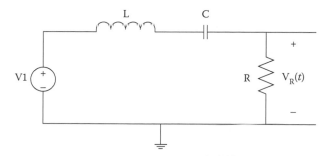

FIGURE 2.5
RLC circuit.

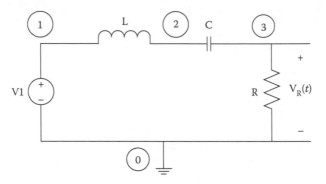

FIGURE 2.6
Figure 2.5 with node numbers and element names.

PSPICE Program

```
RLC circuit
V1 1 0 PULSE(0 5 0.001ms 0.001ms 1ms 1)
L 1 2 2H
C 2 3 1.5e-6
R 3 0 1000
.TRAN   0.2e-3 5e-3
.PRINT TRAN V(3)
.PROBE
.END
```

The results are:

```
TIME V(3)
0.000E + 00 0.000E + 00
2.000E-04  4.711E-01
4.000E-04  8.951E-01
6.000E-04  1.267E + 00
8.000E-04  1.588E + 00
1.000E-03  1.858E + 00
1.200E-03  2.079E + 00
1.400E-03  2.253E + 00
1.600E-03  2.381E + 00
1.800E-03  2.468E + 00
2.000E-03  2.514E + 00
2.200E-03  2.525E + 00
2.400E-03  2.502E + 00
2.600E-03  2.450E + 00
2.800E-03  2.372E + 00
3.000E-03  2.271E + 00
3.200E-03  2.151E + 00
3.400E-03  2.015E + 00
3.600E-03  1.867E + 00
3.800E-03  1.709E + 00
4.000E-03  1.544E + 00
```

```
4.200E-03 1.375E + 00
4.400E-03 1.205E + 00
4.600E-03 1.036E + 00
4.800E-03 8.703E-01
5.000E-03 7.090E-01
```

Initial Conditions

The .**NODESET** command is used to set operating point at specified nodes of a circuit during the initial run of a transient analysis. The general format of the .NODESET command is:

.**NODESET V(node1) = value V(node2) = value2, ...**

where

V(node1), V(node2) are voltages at nodes 1, 2, respectively.

Voltage V(node1) is set to value1, and voltage V(node2) is set to value2, and so on.

.NODESET provides a preliminary guess for voltages at the specified nodes for bias point calculations. The .NODESET command is especially useful for analysis of circuits that have more than one stable state, such as bi-stable circuit. SPICE is guided in calculating the bias point by using the .NODESET command.

The .**IC** statement is only used when the transient analysis statement, .TRAN, includes the "**UIC**" option.

The initial voltage across a capacitor or the initial current flowing through an inductor can be specified as part of a capacitor or inductor component statement. For example, for a capacitor we have:

Cname N+ N−   value IC = initial voltage

and for an inductor, we use the statement

Lname N+ N−   value IC = initial current.

It should be noted that the initial conditions on an inductor or capacitor are used provided .TRAN statement includes the "UIC" option. Example 2.8 illustrates the use of the .IC command.

2.4 AC Analysis

The .AC control statement is used to perform AC analysis on a circuit. The general format of the .AC statement is:

.AC   FREQ_VAR   NP   FSTART   FSTOP

where
> **FREQ_VAR** is one of three keywords that indicates the frequency varia-
> tion by decade (DEC), by octave (OCT), or linearly (LIN);
> **NP** is the number of points, its interpretation depends keyword (DEC,
> OCT, or LIN) word in the FREQ. For:
>> DEC – NP is the number of points per decade
>> OCT – NP is the number of points per octave
>> LIN – NP is the total number of points spaced evenly from frequency
>> FSTART and ending at FSTOP;
> **FSTART** is the starting frequency, FSTART cannot be zero; and
> **FSTOP** is the final or ending frequency.

For example, the statement:

.AC LIN 100 1000 5000

causes AC analysis to be performed with a frequency sweep starting from 1000 Hz and ending at 5000 Hz. The frequency range is divided into 100 equal parts; and 100 evaluations are required in this analysis.
The statement:

.AC DEC 10 100 10000

causes an AC analysis to be done with a frequency sweep from 100 to 100,000 Hz. There are three subintervals (100 to 1000, 1000 to 10,000 and 10,000 to 100,000). Each subinterval has 10 points selected on a logarithmic scale. The following example illustrates .AC command for plotting frequency response.

Example 2.4: Frequency Response of RC Ladder Network

Figure 2.7 is an RC ladder network. If R1 = R2 = R3 = 1 KΩ and C1 = C2 = C3 = 1 μF, plot the magnitude response at the output.

Solution

The PSPICE circuit file program for obtaining the frequency response is:

```
RC Network
VIN 1 0 AC 1 0
R1 1 2 1K
C1 2 0 1uF
R2 2 3 1K
C2 3 0 1uF
R3 3 4 1K
C3 4 0 1uF
.AC DEC 5 10 10000
.PLOT AC VDB(4)  ; plot magnitude in decibels
.PROBE; to plot magnitude response using PROBE
.END
```

The magnitude response is shown in Figure 2.8.

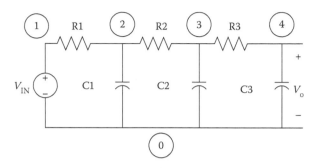

FIGURE 2.7
RC ladder network.

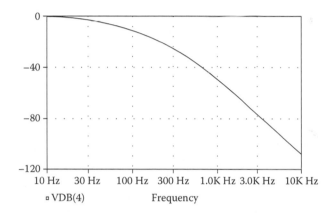

FIGURE 2.8
Magnitude and phase of an RC ladder network.

2.5 Printing and Plotting

The Printing Command

The .PRINT control statement is used to print tabular outputs. The general format of the .PRINT statement is:

.PRINT ANALYSIS_TYPE OUTPUT_VARIABLE

where

ANALYSIS_TYPE can be **DC, AC, TRAN, NOISE,** or **DISTO.** Only one analysis type must be specified for .PRINT statement; and

OUTPUT-VARIABLE can be voltages or currents. Up to eight output variables can accompany one .PRINT statement. If more than eight output variables are to be printed, additional .PRINT statements can be used.

The output variable may be node voltages and current through voltage sources. PSPICE allows one to obtain current flowing through passive elements. The voltage output variable has the general form:

V(node 1, node 2) or V(node 1) if node 2 is node "0."

The current output variable has the general form:

I (Vname)

where
Vname is an independent-voltage source specified in the circuit netlist. For PSPICE, the current output variable can also be specified as I(Rname)

where
Rname is resistance defined in the input circuit.

For example,

.PRINT DC V(4) V(5,6) I(Vsource)

will print DC voltage at node 4, DC voltage between nodes 5 and 6; and the current flowing through an independent voltage source named Vsource. In addition, the statement:

.PRINT TRAN V(1) V(7,3)

will print the voltage at node 1 and voltage between nodes 7 and 3 for a transient analysis.

For AC analysis, output voltage and current variables may be specified as magnitude, phase, real, or imaginary. Table 2.2 shows the name types for AC output variables.

For example,

.PRINT AC VDB(3) VP(3)

will plot the voltage magnitude in dB and phase in degrees of voltage at node 3.

TABLE 2.2

Name Types for AC Output Variable

Output Variable	Meaning
V OR I	Magnitude of V or I
VR or IR	Real part of complex value V or I
VI or II	Imaginary part of complex number V or I
VM or IM	Magnitude of complex number V or I
VDB or IDB	Decibel value of magnitude, i.e., $20 \log_{10} |V|$ or $20 \log_{10} |I|$
VG or IG	Group delay of complex number

The Plot Command

The .PLOT control statement is used to generate line printer plots of specified output variables. The general format of the .PLOT command is

.PLOT ANALYSIS_TYPE OUTPUT_VARIABLE PLOT_LIMITS

where the

ANALYSIS_TYPE can be **DC, AC, TRAN, NOISE, or DISTO.** Only one analysis type must be specified for .PRINT statement;

OUTPUT_VARIABLE can be voltages or currents. The methods of specifying the output variables are similar to those in the .PRINT control statement described in Section 2.5; and

PLOT_LIMITS specifies the lower and upper limit values that should appear on y-axis for a specified output variable. The PLOT_LIMITS may be omitted, in that case PSPICE assigns a plotting range of the specified output variable. The plot limits should come immediately after the output variable where the plotting range corresponds.

For example,

.PLOT DC V(4,3) I(VIN)

will plot the DC values of the voltage between nodes 4 and 3. In addition, the current through the independent voltage source VIN will be plotted. No plotting range is specified, so PSPICE will assign a default range for plotting y-axis. The range and increment of the x-axis should be specified in the .DC control statement. In addition, the statement

.PLOT AC VDB(5)(0,60)

will plot the voltage magnitude at node 5 between the range of 0 to 60 dB.

2.6 Transfer Function Command

The .TF command can be used to obtain the small-signal gain, DC input resistance, and DC output resistance of a circuit by linearizing the circuit around a bias point. The format for the .TF command is:

.TF OUTPUT_VARIABLE INPUT_SOURCE

where

OUTPUT_VARIABLE can be voltage or current. If the OUTPUT_VARIABLE is a current, it is restricted to current flowing through a voltage source; and

BOX 2.1   STEPS FOR PERFORMING ORCAD TRANSFER FUNCTION ANALYSIS

- Draw the circuit using the steps of Box 1.2.
- Select "PSPICE/New Simulation Profile."
- Insert the name of the simulation in "New Simulation" menu.
- Click on "Create."
- In the "Simulation Settings," select "Bias Point" for analysis type, and choose "General Settings" under option.
- Under "output file options," check "calculate small-signal DC Gain (.TF)."
- Insert names for "the input source name," and a variable "to output variable."
- Select "OK" to close "Simulation Settings" dialog box.
- To run the DC analysis, choose "PSPICE/Run."
- View the simulation results either by choosing "View/Output" for the output text file or by going to ORCAD Capture and selecting "V" to view the nodal voltages.

INPUT_SOURCE must be an independent voltage or current source. If the input source is current source, then a large resistance must be connected in parallel with the current source.

The SPICE output file contains the following information:

- The ratio of output_variable/input_source
- The DC input resistance with respect to INPUT_SOURCE
- The DC output resistance with respect to OUTPUT_VARIABLE.

The above three output statements are printed irrespective of the existence of .PRINT, .PLOT, or .PROBE statements in the SPICE circuit file program.

ORCAD schematics can also be used to obtain the transfer function. The circuit is drawn by use of Box 1.2. The DC bias analysis is used. However, under "output file options," check "calculate small-signal DC Gain" (.TF). Insert names at "the input source name," and "To output variable." The input source and output variable should be currents or voltages. The steps for performing the transfer function analysis are shown in Box 2.1.

The following example shows the application of the .TF command.

Example 2.5: Input and Output Resistance of Resistive Network

A resistive network is shown in Figure 2.9. If the values of all the resistors are 10 Ω, find the input resistance R_{IN}, and output resistance R_{OUT}.

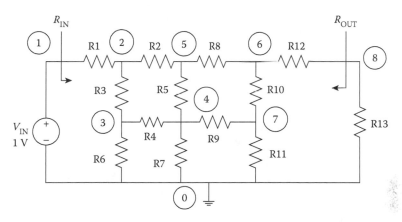

FIGURE 2.9
Resistive network.

Solution

PSPICE Program

```
RESISTIVE NETWORK
VIN     1       0       DC      1
R1      1       2       10
R2      2       5       10
R3      2       3       10
R4      3       4       10
R5      5       4       10
R6      3       0       10
R7      4       0       10
R8      5       6       10
R9      4       7       10
R10     6       7       10
R11     7       0       10
R12     6       8       10
R13     8       0       10
.TF     V(8)    VIN
.END
```

Using the steps in Box 2.1, the circuit can be drawn and simulated. The simulation settings for Example 2.5 is shown in Figure 2.10.
PSPICE results are:

```
**** SMALL-SIGNAL CHARACTERISTICS
V(8)/VIN = 6.977E-02
INPUT RESISTANCE AT VIN = 1.955E + 01
OUTPUT RESISTANCE AT V(8)  = 6.583E + 00
```

From the results, the input DC resistance is 1.955E + 01 Ω and output DC resistance is 6.583E + 00 Ω.

The .TF command can be used to obtain the Thevenin equivalent circuit of a complex circuit. The Thevenin resistance can be obtained by specifying the

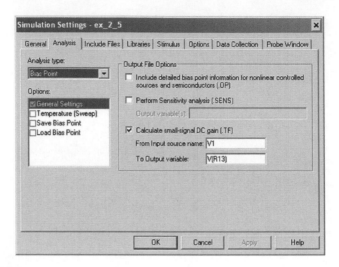

FIGURE 2.10
Simulation settings for performing transfer function analysis.

output-variable in .TF command as voltage between the specified nodes. The Thevenin voltage is obtained from the data of the node voltages obtained from the PSPICE output. The following example illustrates this application of the .TF command for obtaining a Thevenin equivalent circuit.

Example 2.6: Thevenin Equivalent Circuit of a Network

In Figure 2.11, R1 = 4 KΩ, R2 = 8 KΩ, R3 = 10 KΩ, R4 = 2 KΩ, R5 = 8 KΩ, R6 = 6 KΩ, and V1 = 10 V. If the current source I1 is 5 mA, find the Thevenin equivalent circuit to the left of nodes A and B of the circuit. In addition, find the power dissipated in a 2 KΩ resistor that is connected between nodes A and B.

Solution

Figure 2.12 is basically Figure 2.11 with node numbers.
The PSPICE circuit file program for obtaining the equivalent circuit is as follows.

PSPICE Program

```
THEVENIN EQUIVALENT CIRCUIT
V1      1       0       DC 10V
R1      1       2       4K
R2      2       0       8K
R3      2       3       10K
R4      3       0       2K
R5      2       4       8K
R6      4       0       6K
I1      3       4       5MA
.TF     V(4)    V1
.END
```

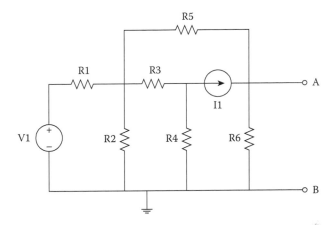

FIGURE 2.11
Circuit for Example 2.6.

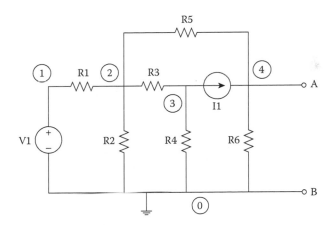

FIGURE 2.12
Figure 2.10 with node numbers.

The relevant data from PSPICE output is:

```
NODE VOLTAGE NODE VOLTAGE NODE VOLTAGE NODE VOLTAGE
( 1) 10.0000 (2) 7.1910 (3) -7.1348 (4) 20.2250
**** SMALL-SIGNAL CHARACTERISTICS
V(4)/V1 = 2.022E-01
INPUT RESISTANCE AT V1 = 7.574E + 03
OUTPUT RESISTANCE AT V(4) = 3.775E + 03
```

The Thevenin resistance R_{TH} is 3.775E + 03 Ω.
The Thevenin voltage V_{TH} is the voltage at node 4, which is 20.2250 Volts. The Thevenin equivalent circuit is drawn in Figure 2.13.

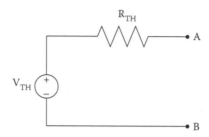

FIGURE 2.13
Thevenin equivalent circuit of Figure 2.10.

The power dissipated in 2 KΩ resistor is:

$$P = \left(\frac{V_{TH}}{2000 + R_{TH}}\right)^2 2000 = 0.0245 \text{ Watts}$$

2.7 DC Sensitivity Analysis

The DC sensitivity of circuit element values and variation of model parameters on selected output variables is obtained using the .SENS statement. The general format for using the .SENS command is:

.SENS   OUTPUT_VARIABLE

where
OUTPUT_VARIABLE can be voltage or current. If the **OUTPUT_VARIABLE** is a current, it is restricted to current flowing through a voltage source.

The circuit under consideration is linearized about the bias point and the sensitivities of each output variable to all the element values and model parameters are calculated. If there are the following elements in a circuit (R_1, R_2, R_3, and V_{S1}) and the output variable of interest is V_X, then

$$V_X = f(R_1, R_2, R_3, V_{S1})$$

BOX 2.2   STEPS FOR PERFORMING ORCAD SCHEMATIC SENSITIVITY ANALYSIS

- Draw the circuit using the steps of Box 1.2.
- Select "PSPICE/New Simulation Profile."
- Insert the name of the simulation in "New Simulation" menu.
- Click on "Create."
- In the "Simulation Settings," select "Bias Point" for analysis type, and choose "General Settings" under option.
- Under "output file options," check "Perform Sensitivity Analysis (.SENS)."
- Insert variable names "Output variable(s)."
- Select "OK" to close "Simulation Settings" dialog box.
- To run the DC analysis, choose "PSPICE/Run."
- View the simulation results by choosing "View/Output."

The .SENS command will give the

$$\frac{\partial V_X}{\partial R_i}, \frac{\partial V_X}{\partial R_i}\left(\frac{R_i}{100}\right) \text{ where } R_i = 1, 2, \text{and } 3$$

$$\frac{\partial V_X}{\partial V_{S1}}, \frac{\partial V_X}{\partial V_{S1}}\left(\frac{V_{S1}}{100}\right).$$

Both the absolute sensitivity, $\partial V_X/\partial R_i$ or $\partial V_X/\partial V_{S1}$ and the relative sensitivity, $\partial V_X/\partial R_i$ ($R_i/100$) or $\partial V_X/\partial V_{S1}$ ($V_{S1}/100$) will be outputted.

In using PSPICE schematic for the sensitivity analysis, the schematic should be captured using the steps in Table 1.2. In the schematic, we can include "off-page connector" that allows you to label a node in the circuit and refer to that node by name. We use the "bias Point" analysis to perform the sensitivity analysis. In the "output file options," check the box labeled "Perform Sensitivity Analysis (.SENS). Enter the variable names in the text box labeled "output variables." The output variables indicate the circuit variables to be monitored during simulation. The steps for performing the sensitivity analysis are summarized in Box 2.2.

The following example illustrates the use of the .SENS statement.

Example 2.7: DC Sensitivity of a Bridge-T Network

In the bridge-T network shown in Figure 2.14, R1 = 20 KΩ, R2 = 40 KΩ, R3 = 20 KΩ, R4 = 50 KΩ, R5 = 10 KΩ, and VS = 10 V. Use PSPICE to compute the sensitivity of the voltage across the resistor R5 with respect to the circuit elements.

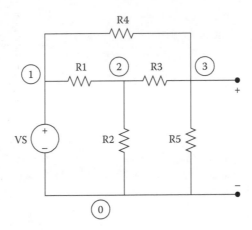

FIGURE 2.14
Bridge-T network.

Solution

PSPICE Program

```
BRIDGE-T NETWORK
VS 1 0 DC 10V
R1 1 2 20K
R2 2 0 40K
R3 2 3 20K
R4 1 3 50K
R5 3 0 10K
.SENS V(3)
.END
```

Using the steps in Box 2.2, the circuit can be drawn and simulated. The simulation settings for Example 2.7 is shown in Figure 2.15.

The relevant PSPICE results are:

```
**********************************************************
BRIDGE-T NETWORK
 DC SENSITIVITY ANALYSIS TEMPERATURE = 27.000 DEG C
**********************************************************
DC SENSITIVITIES OF OUTPUT V(3)
 ELEMENT ELEMENT ELEMENT NORMALIZED
 NAME VALUE SENSITIVITY SENSITIVITY
 (VOLTS/UNIT) (VOLTS/PERCENT)
 R1 2.000E + 04 -3.289E-05 -6.578E-03
 R2 4.000E + 04 8.444E-06 3.378E-03
 R3 2.000E + 04 -2.400E-05 -4.800E-03
 R4 5.000E + 04 -1.956E-05 -9.778E-03
 R5 1.000E + 04 1.778E-04 1.778E-02
 VS 1.000E + 01 2.667E-01 2.667E-02
**********************************************************
```

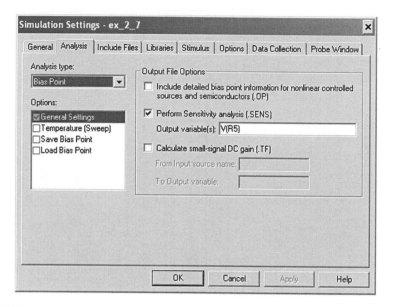

FIGURE 2.15
Simulation setting for sensitivity analysis.

The four-column tabular output has the following headers: ELEMENT NAME, ELEMENT VALUE, ELEMENT SENSITIVITY, and NORMALIZED SENSITIVITY. The ELEMENT SENSITIVITY is the absolute sensitivity in amperes or volts per unit of the respective element. The NORMALIZED SENSITIVITY is amperes or volts per 1% variation in the value of the respective element. The most informative data are the normalized sensitivities.

In the circuit, 1% change in R1 causes about 6.6 mV variation in the voltage at node 3. 1% change in VS causes roughly 26.7 mV variation in the voltage at node 3. The largest variation in the voltage at node 3 is caused by 1% change in VS. The smallest variation in the output voltage is caused by 1% change in R2. An increase in R1, R3, and R4 causes the output voltage at node 3 to decrease, whereas an increase in R2, R5, and VS brings about an increase in the output voltage.

Example 2.8: Transient Analysis of a Sequential Circuit

For the sequential circuit shown in Figure 2.16, find the voltages across the 50 Ω resistor when the switch moved from a to b at $t \geq 0$.

Solution

At $t < 0$, the voltage across the capacitor is:

$$V_C(0) = \frac{(20)*(300)}{400} = 15V.$$

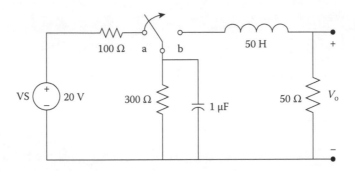

FIGURE 2.16
RLC circuit for Example 2.8.

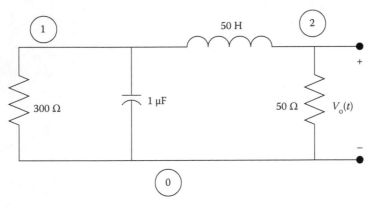

FIGURE 2.17
Equivalent circuit of Figure 2.16.

The current through the inductor is $I_L(0) = 0A$.

The equivalent circuit of Figure 2.16 for $t > 0$ is shown in Figure 2.17.

The PSPICE circuit file program for obtaining the voltage $v_o(t)$ is:

PSPICE Program

```
RLC circuit
R1      1       0       300
C1      1       0       1uF     IC = 15V
L1      1       2       50      IC = 0A
R2      2       0       50
.TRAN   0.01    0.5     UIC
.PLOT   TRAN    V(2)
.PROBE  ; TO PLOT VOLTAGE AT NODE 2
.END
```

The output is plotted in Figure 2.18.

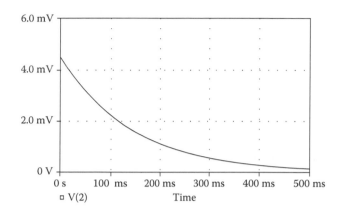

FIGURE 2.18
Output voltage $v_o(t)$.

2.8 Temperature Analysis

All elements in a SPICE netlist are assumed to be measured at the nominal temperature, TNOM, of 27°C (300°K). The nominal temperature of 27°C can be changed by using .OPTIONS command. All simulations are performed at the nominal temperature. The .TEMP command is used to change the temperature at which a simulation is performed. The general format of the .TEMP statement is:

.TEMP TEMP1 TEMP2 TEMP3 ... TEMPN

where
 TEMP1, TEMP2, and **TEMP3** are temperatures at which the simulations are performed.

For example, the statement:

.TEMP 120 200

is the command to perform circuit calculations at 120°C and 200°C.

 When a temperature is changed, elements such as resistors, capacitors, and inductors have values that may change. In addition, adjustments are made to devices, such as transistors and diodes, whose models are temperature dependent.

If one desires to sweep the circuit temperature over a range of values, the .DC sweep command can be used. The general syntax for this sweep is:

.DC TEMP START_VALUE STOP_VALUE INCREMENT

where
   START_VALUE is the starting temperature in °C;
   STOP_VALUE is the ending value of temperature; and
   INCREMENT is the step size.

For example, the statement:

.DC TEMP 0 100 10

will make SPICE calculate all parameters of the circuit being analyzed at starting temperature of 0°C and ending the simulation at 100°C. The increment for the analysis is 10°C.

Since DC sweep can be nested, it is possible to sweep a component and a source while still sweeping temperature.

2.9 PROBE Statement

The general format for specifying PROBE statements is:

.PROBE OUTPUT_VARIABLES

where
   OUTPUT_VARIABLES can be node voltages and/or device currents. If no OUTPUT_VARIABLE is specified, PROBE will save all node voltages and device currents.

PSPICE creates a data file, probe.dat, for use by PROBE. The file is used by PROBE to display simulation results in graphical format. When PROBE is invoked, there are several commands available in the main menu for file accessing, plotting, editing, viewing, and adding or removing trace. PSPICE Reference manual should be consulted for details.

PSPICE has several functions that PROBE can use to determine various characteristics of a circuit from variables available in PROBE. The functions can be found in Table 1.4.

Example 2.9: Power Calculations of an RL Circuit Using PROBE

For the RL circuit shown in Figure 2.19, $v(t) = 10 \sin(200\pi t)$ volts. Use PROBE to plot the average power P_{AVE} delivered to the resistor R as a function of time.

Solution

The input voltage is a sinusoidal voltage, we use the sinusoidal waveform included as a function of the transient analysis sources, given as

$$\text{Sin}(V_O, V_a, V_{freq}, \text{ td, df, phase}).$$

In this problem,

$$V_O = 0 \text{ V}; \qquad V_a = 10 \text{ V}; \qquad freq = 100 \text{ Hz}$$

$$td = 0 \text{ second}; \qquad df = 0; \qquad phase = 0.$$

Since the analysis has time as an independent variable, transient analysis is performed.

PSPICE Program

```
RL CIRCUIT AND PROBE
VS      1       0       SIN(0 10 100 0 0 0)
L1      1       2       1MH
R1      2       1       100
*CONTROL STATEMENTS
.TRAN   1.0E-3 3.0E-2
.PROBE
.END
```

After running the program and invoking PROBE, the PROBE expression that can be used to obtain the average power is:

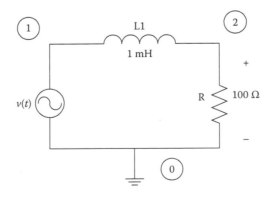

FIGURE 2.19
RL circuit.

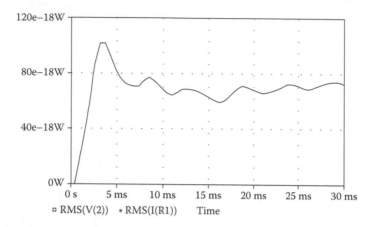

FIGURE 2.20
Average power from PROBE plot.

$$\text{Average Power} = rms\ (V(2))^*\ rms(I(R1)).$$

The graphic display obtained from PROBE plot is shown in Figure 2.20.

Example 2.10: Input Impedance versus Frequency of a Filter Network

For the passive filter network shown in Figure 2.21, R1 = R2 = R3 = 500 Ω, R4 = 1000 Ω, C1 = C2 = C3 = 1.5 μF, L1 = 2 mH, L2 = 4 mH, and L3 = 6 mH. Use PROBE to find the input impedance $|Z_{IN}(w)|$ with respect to frequency.

Solution

The PSPICE circuit file program for analyzing the circuit is:

PSPICE Program

```
FILTER CIRCUIT
VS      1      0      AC      1      0
R1      1      2      500
L1      2      5      2E-3
C1      5      0      1.5E-6
R2      2      3      500
L2      3      6      4E-3
C2      6      0      1.5E-6
R3      3      4      500
L3      4      7      6E-3
C3      7      0      1.5E-6
R4      4      0      1000
*  CONTROL  STATEMENTS
.AC      DEC      10      1.0E2   1.0E7
.PROBE
.END
```

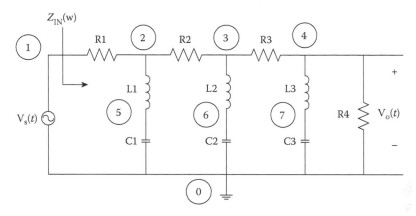

FIGURE 2.21
Passive filter network.

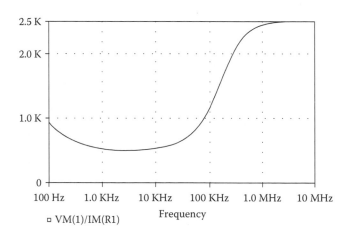

□ VM(1)/IM(R1)   Frequency

FIGURE 2.22
Input impedance versus frequency.

The PROBE expression for obtaining the input impedance as a function of frequency is:

$$Z_{IN}(w) = VM(1) / IM(R1)$$

The plot of the input impedance is shown in Figure 2.22.

Problems

2.1   For Figure 2.2, R1 = R2 = R3 = 100 Ω, R4 = 400 Ω, R5 = 500 Ω, and R6 = 50 Ω. Determine the current IB if the source voltage VS = 10 V.

2.2   For Figure P2.2, L = 2 H and R = 400 Ω. If $V_s(t) = 10\exp(-12t)$ cos(1000πt) with a duration of 2 ms, find the output waveform $V_o(t)$.

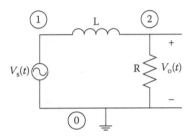

FIGURE P2.2
RL circuit.

2.3   Plot the magnitude response of the Wein-Bridge circuit shown in Figure P2.3. Assume that C1 = C2 = 4 nF, R1 = R3 = R4 = 5 KΩ, and R2 = 10 KΩ. (a) What is the center frequency? (b) What is the bandwidth?

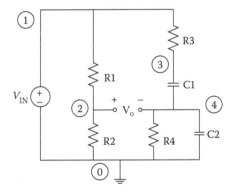

FIGURE P2.3
Wein-Bridge circuit.

2.4   The simplified equivalent circuit of an amplifier is shown in Figure P2.4. Use SPICE to obtain the input and output resistance. What is voltage gain at DC? Assume that RGS = 100 KΩ, Rds = 50 KΩ, RS = 50 Ω, RL = 10 KΩ, and RLC = 5 KΩ.

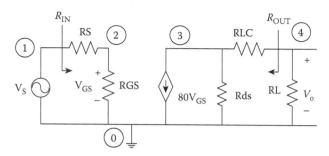

FIGURE P2.4
Simplified equivalent circuit of an amplifier.

2.5 For the multi-source resistive circuit shown in Figure P2.5, $I1 = 2$ mA, $V1 = 5$ V, $V2 = 4$ V, $R1 = 1$ KΩ, $R2 = 4$ KΩ, $R3 = 2$ KΩ, $R4 = 10$ KΩ, $R5 = 8$ KΩ, $R6 = 7$ KΩ, and $R7 = 4$ KΩ. Find the Thevenin equivalent circuit at nodes A and B.

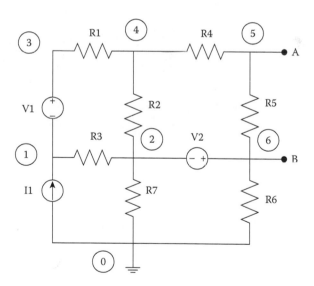

FIGURE P2.5
Multi-source resistive circuit.

2.6 Compute the sensitivity of the output voltage with respect to the circuit elements. Assume that $V1 = 10$ V, $R2 = R3 = 4$ KΩ, $R4 = R5 = 8$ KΩ, and $R1 = R6 = 2$ KΩ.

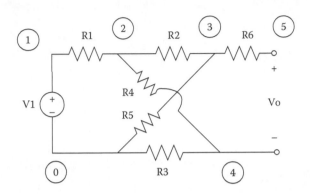

FIGURE P2.6
Resistive circuit.

2.7   For the RLC circuit shown in Figure P2.7, V1 = 8 V, R1 = 100 Ω, R2 = 400 Ω, L1 = 5 mH, and C1 = 20 μF. The switch moves from point A to B at $t = 0$. Find the circuit $i(t)$ after $t > 0$.

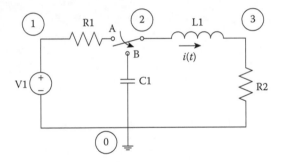

FIGURE P2.7
RLC circuit.

2.8   For the multistage RC circuit shown in Figure P2.8, C1 = C2 = C3 = 1 μF and R1 = R2 = R3 = 1 KΩ. If the input signal $V_s(t) = 20 \cos(120\pi t + 30°)$ V, (a) determine the voltage $V_o(t)$ and (b) find the average power dissipated by R3.

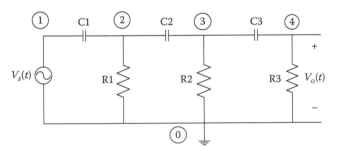

FIGURE P2.8
Multistage RC network.

2.9 For circuit shown in Figure P2.9a, R1 = 300 Ω and R2 = 200 Ω. The input signal is a triangular wave shown in Figure P2.9b. Use PROBE to plot instantaneous voltage and the rms voltage across the 200 Ω resistor. What is the ratio of the rms voltage to the average voltage of the triangular wave?

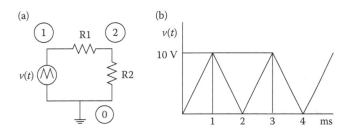

FIGURE P2.9
(a) Resistive circuit and (b) input waveform.

2.10 The circuit shown in Figure P2.10 has two sources with the same amplitude and frequency but different phases. R1 = R2 = 100 Ω, R3 = 5 KΩ, L1 = L2 = 1 mH, VS1(t) = 168sin(120πt) V and VS2(t) = 168sin(120πt + 60°) V. Use PROBE to determine the average power supplied to or obtained from the sources VS1 and VS2. In addition, determine the power supplied to resistor R3.

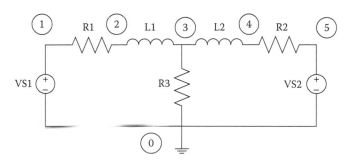

FIGURE P2.10
Circuit with two sources.

2.11 For the twin-T network shown in Figure P2.11, R1 = R4 = 1 KΩ, R2 = R3 = 2 KΩ, R5 = 5 KΩ, C1 = C2 = 1 μF, and C3 = 0.5 μF. Use PROBE to plot the magnitude of the input impedance Z_{IN} with respect to frequency of source VS.

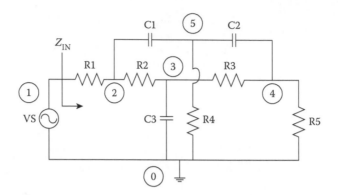

FIGURE P2.11
Twin-T circuit.

2.12 For the circuit shown in Figure P2.12, find the Thevenin equivalent circuit at nodes A and B. The large resistance R6 has been
connected across nodes A and B to prevent floating nodes in the
circuit.

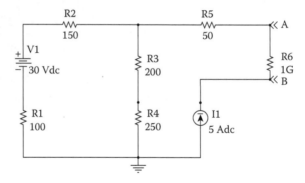

FIGURE P2.12
Passive circuit for obtaining equivalent circuit.

2.13 Compute the sensitivity of the output voltage, V_{OUT}, with respect
to the circuit elements.

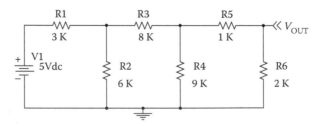

FIGURE P2.13
Circuit for sensitivity analysis.

2.14 Use PSPICE to plot the magnitude response of the voltage across the 800 Ω resistor. Determine the center frequency and the bandwidth.

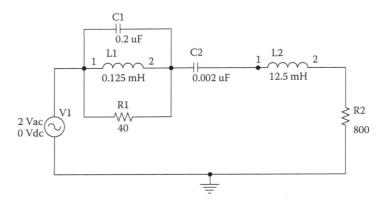

FIGURE P2.14
RLC circuit for frequency response analysis.

Bibliography

1. Al-Hashimi, Bashir. *The Art of Simulation Using PSPICE, Analog, and Digital*. Boca Raton, FL: CRC Press, 1994.
2. Ellis, George. "Use SPICE to Analyze Component Variations in Circuit Design," In *Electronic Design News (EDN)*, (April 1993): 109–14.
3. Eslami, Mansour, and Richard S. Marleau. "Theory of Sensitivity of Network: A Tutorial." *IEEE Transactions on Education*, Vol. 32, no. 3 (August 1989): 319–34.
4. Fenical, L. H. *PSPICE: A Tutorial*. Upper Saddle River, NJ: Prentice Hall, 1992.
5. Kavanaugh, Micheal F. "Including the Effects of Component Tolerances in the Teaching of Courses in Introductory Circuit Design." *IEEE Transactions on Education*, Vol. 38, no. 4 (November 1995): 361–64.
6. Keown, John. *PSPICE and Circuit Analysis*. New York: Maxwell Macmillan International Publishing Group, 1991.
7. Kielkowski, Ron M. *Inside SPICE, Overcoming the Obstacles of Circuit Simulation*. New York: McGraw-Hill, Inc., 1994.
8. Nilsson, James W., and Susan A. Riedel. *Introduction to PSPICE Manuel Using ORCAD Release 9.2 to Accompany Electric Circuits*. Upper Saddle River, NJ: Pearson/Prentice Hall, 2005.
9. OrCAD Family Release 9.2. San Jose, CA: Cadence Design Systems, 1986–1999.
10. Rashid, Mohammad H. *Introduction to PSPICE Using OrCAD for Circuits and Electronics*. Upper Saddle River, NJ: Pearson/Prentice Hall, 2004.
11. Spence, Robert, and Randeep S. Soin. *Tolerance Design of Electronic Circuits*. London: Imperial College Press, 1997.
12. Soda, Kenneth J. "Flattening the Learning Curve for ORCAD-CADENCE PSPICE," *Computers in Education Journal*, Vol. XIV (April–June 2004): 24–36.

13. Svoboda, James A. *PSPICE for Linear Circuits*. 2nd ed. New York: John Wiley & Sons, Inc., 2007.
14. Tobin, Paul. "The Role of PSPICE in the Engineering Teaching Environment." Proceedings of International Conference on Engineering Education, Coimbra, Portugal, September 3–7, 2007.
15. Tobin, Paul. *PSPICE for Circuit Theory and Electronic Devices*. San Rafael, CA: Morgan & Claypool Publishers, 2007.
16. Tront, Joseph G. *PSPICE for Basic Circuit Analysis*. New York: McGraw-Hill, 2004.

3

PSPICE Advanced Features

This chapter is a continuation of the discussion of the features of PSPICE. Several of the PSPICE control statements that were not discussed in Chapter 2 are covered in this chapter. We briefly discuss the device models. This is followed by methodologies for changing component values. Subcircuit is defined and subcircuit calls are discussed. Analog behavior model and Monte Carlo analysis are also presented.

3.1 Device Model

The .MODEL statement specifies a set of device parameters that can be referenced by elements or devices in a circuit. The general form of the .MODEL statement is

MODEL MODEL_NAME MODEL TYPE PARAMETER_NAME = VALUE

where

MODEL_NAME is a name for which devices use to reference a particular model. The model name must start with a letter. To avoid confusion, it is advisable to make the first character of the model_name identical with the first character of the device name. See Table 1.1 for a list of the element names;

MODEL_TYPE refers to the device type, which can be active or passive. The MODEL_TYPEs available in PSPICE are shown in Table 3.1. The reference model may be available in the main circuit file, or accessed through a .INC statement, or may be in a library file. A device cannot reference a model statement that does not correspond to that type of model. It is possible to have more than one model of the same type in the circuit file, but they must have different model names; and

PARAMETER_VALUES follow the model type. The model parameter values are enclosed in parenthesis. It is not required to list all the parameters values of the device. Parameters not specified are assigned default values.

In general, the .MODEL statement should adhere to the following rules:

1. More than one .MODEL statement can appear in a circuit file and each .MODEL statement should have a different model name. For example, the following models of a MOSFET are valid.

$$M1 \quad 1 \quad 2 \quad 0 \quad 0 \quad MOD1 \quad L = 5U \quad W = 10U$$

$$M2 \quad 1 \quad 2 \quad 0 \quad 0 \quad MOD2 \quad L = 5U \quad W = 15U$$

$$.MODEL \quad MOD1 \quad NMOS \quad (VTO = 1.0 \; KP = 200)$$

$$.MODEL \quad MOD2 \quad NMOS \quad (VTO = 1.0 \; KP = 150)$$

There are two model names, MOD1, MOD2 for model type NMOS.

2. More than one device of the same type may reference a given model using the .MODEL statement. For example,

$$D1 \quad 1 \quad 2 \quad DMOD$$

$$D2 \quad 2 \quad 3 \quad DMOD$$

$$.MODEL \; DMOD \; D \; (IS = 1.0E\text{-}14 \; CJP = 0.3P \; VJ = 0.5)$$

Two diodes, D1 and D2, reference the given diode model DMOD.

TABLE 3.1

Model Types of Devices

Type of Device	Model Type	Recommended Instance Name
Capacitor	CAP	CXXX
Inductor	IND	LXXX
Resistor	RES	RXXX
Diode	D	DXXX
NPN Bipolar transistor	NPN	QXXX
PNP Bipolar transistor	PNP	QXXX
Lateral PNP bipolar transistor	LPNP	QXXX
N-channel junction FET	NJF	JXXX
P-channel junction FET	PJF	JXXX
N-channel MOSFET	NMOS	MXXX
P-channel MOSFET	PMOS	MXXX
N-channel GaAs MESFET	GASFET	BXXX
Nonlinear, magnetic curve (transformer)	CORE	KXXX
Voltage-controlled switch	VSWITCH	SXXX
Current-controlled switch	ISWITCH	WXXX

3. A device cannot reference a model statement that does not correspond to the device. For example, the following statements are incorrect.

R1   1   2   DMOD

.MODEL DMOD D (IS = 1.0E-12)

Resistor R1 *cannot* reference a diode model DMOD

Q1   3   2   1   MMDEL

.MODEL MMDEL NMOS (VTO = 1.2)

The bipolar transistor cannot reference the NMOS transistor.

4. A device cannot reference more than one model in a netlist.

In the following sections, we shall discuss the .MODEL statements for both passive (R, L, C) and active elements (D, M, Q).

3.1.1 Resistor Models

In Section 2.1.1, the basic description of passive elements was expressed in terms of element name, nodal connections, and component value. To model a resistor, two statements are required. The general format is:

Rname   NODE1   NODE2   MODEL_NAME   R_VALUE

.MODEL   MODEL_NAME   RES[MODEL_PARAMETER]

where
   MODEL_NAME is a name preferably starting with the character R. It can
      be up to eight characters long;
   RES is the specification for PSPICE model type associated with resistors;
      and
   MODEL_PARAMETERS are parameters that can vary. Table 3.2 shows
      the model parameters for resistors and their default values.

PSPICE uses the model parameters to calculate the resistance using the following equations:

$$\text{R\_model(T)} = R\_value^* \, R[1 + TC1(T - T_{nom}) + TC2(T - T_{nom})^2] \quad \textbf{(3.1)}$$

or

$$\text{R\_model} = R\_value^* R[1.01]^{TCE(T-Tnom)} \quad \textbf{(3.2)}$$

TABLE 3.2

Resistor Model Parameters and their Default Values

Model Parameter	Description	Default Value	Unit
R	Resistance multiplier	1	
TC1	Linear temperature coefficient	0	°C^{-1}
TC2	Quadratic temperature coefficient	0	°C^{-2}
TCE	Exponential temperature coefficient	Default value	%C

where

T is the temperature at which the resistance needs to be calculated.

Equation 3.1 uses the linear and quadratic temperature coefficients of the resistor. The coefficients TC1 and TC2 are specified in the .MODEL statement. Equation 3.2 is used if the resistance is exponentially dependent on temperature.

For example, the statements:

$$R1 \quad 4 \quad 3 \quad RMOD1 \quad 5K$$

$$.MODEL \quad RMOD1 \quad RES(R = 1, \quad TC1 = 0.00010)$$

describe a resistor with model name RMOD1. The .MODEL statement specifies that the resistor R1 has a linear temperature coefficient of + 100 ppm/°C.

The statement:

$$R2 \quad 5 \quad 4 \quad RMOD2 \quad 10K$$

$$.MODEL \quad RMOD2 \quad RES(R = 2, \quad TCE = 0.0010)$$

describes a resistor whose value with respect to temperature is given by the expression

$$R2(T) = 10,000 * 2 * (1.01)^{0.001(T - Tnom)} \tag{3.3}$$

where

$Tnom$ is the normal temperature that can be set with $Tnom$ option.

3.1.2 Capacitor Models

Whereas the resistors were modeled to be temperature dependent, capacitors can be both temperature and voltage dependent. In addition, capacitors can have initial voltage impressed on them. The element description and the model statements have the feature to incorporate the above influences on capacitors. The general format for modeling capacitors is:

CNAME NODE + NODE- MODEL_NAME VALUE IC = INITIAL_VALUE

.MODEL   MODEL_NAME CAP MODEL_PARAMETERS

where
   MODEL_NAME is a name (preferably starting with character C). It can be up to eight characters long;
   CAP is the PSPICE specification for the model type associated with capacitors; and
   MODEL_PARAMETERS are parameters that can be used to describe the capacitance value with respect to changes in temperature and voltage. Table 3.3 shows the capacitor model parameters.

PSPICE uses the model parameters to calculate the capacitance at a particular temperature, T, and voltage, V, using the following expression:

$$C(V,T) = C\_value*C[1 + VC1*V + VC2*V^2]*$$

$$[1 + TC1(T - T_{nom}) + TC2(T - T_{nom})^2]$$

(3.4)

where
   T_{nom} is nominal temperature set by T_{nom} option.
   For example, the statements:

$$\textbf{CBIAS   5   0   CMODEL   20e-6   IC = 3.0}$$

$$\textbf{.MODEL   CMODEL   CAP(C = 1, VC1 = 0.0001,}$$
$$\textbf{VC2 = 0.00001, TC1 = -0.000006)}$$

describe a capacitance that is a function of both voltage (V) and temperature (T) and whose value is given as

$$C(V,T) = 20.0*10^{-6}[1 + 0.001V + 0.00001V^2]*[1 - 0.000006(T - T_{nom})] \quad (3.5)$$

TABLE 3.3

Capacitor Model Parameters

Model Parameter	Description	Default Value	Unit
C	Capacitance multiplier	1	–
TC1	Linear temperature coefficient	0	C^{-1}
TC2	Quadratic temperature coefficient	0	C^{-2}
VC1	Linear voltage coefficient	0	V^{-1}
VC2	Quadratic voltage coefficient	0	V^{-2}

3.1.3 Inductor Models

Inductors are current and temperature independent. They are thus modeled similarly to capacitors. The general format for modeling inductor is:

LNAME NODE + NODE- MODEL_NAME VALUE IC = INITIAL VALUE

.MODEL MODEL_NAME IND MODEL_PARAMETER

where

MODEL_NAME is a name (preferably starting with L). It can be up to eight characters long;

IND is the PSPICE specification for model type associated with inductors; and

MODEL_PARAMETERS are parameters that can be used to express the inductance value as a function of temperature and current. Table 3.4 shows the inductor model parameters.

The inductance at a particular current and temperature is given by the expression:

$$L(I, T) = L_{value} * L[1 + IL1*I + IL2*I^2][1 + TC1(T - T_{nom}) + TC2(T - T_{nom})^2] \quad \textbf{(3.6)}$$

where

T_{nom} in nominal temperature is set by T_{nom} option.

For example, the statements:

L1   6   5   LMOD   25m   IC = 1.5A

.MODEL   LMOD   IND(L = 1   IL1 = 0.001   TC1 = -0.00002)

describe an inductor of value 25 millihenries with initial current of 1.5 A whose value is a function of both current (*I*) and temperature (*T*), and it is given as

$$L(I,T) = 25.0 * 10^{-3}[1 + 0.001I] * [1 - 0.00002(T - T_{nom})]. \quad \textbf{(3.7)}$$

TABLE 3.4

Inductor Model Parameters

Model Parameters	Description	Default Value	Unit
L	Inductance multiplier	1	–
TC1	Linear temperature coefficient	0	°C^{-1}
TC2	Quadratic temperature coefficient	0	°C^{-2}
IL1	Linear current coefficient	0	A^{-1}
IL2	Quadratic current coefficient	0	A^{-2}

The following example explores the temperature effects on a notch filter.

Example 3.1: Temperature Effects on Notch Filter

A notch filter circuit is shown in Figure. 3.1, find the change in the notch frequency as the temperature increases from 25°C to 100°C. Assume the following values for TC1 and TC2:

    for R, TC1 = 1.0E-5 and TC2 = 0
    for C, TC1 = −6.06E-6 and TC2 = 0
    for L, TC1 = 1.0E-7 and TC2 = 0

Solution

PSPICE Netlist

```
NOTCH FILTER AND TEMPERATURE
.OPTIONS RELTOL = 1E-8
.OPTIONS NUMDGT = 6
VS   1   0    AC  1   0
R1   1   2    RMOD   1K
.MODEL   RMOD   RES(R = 1 TC1 = 1.0E-7)
L1   2   3    LMOD   10wE-6
.MODEL   LMOD   IND(L = 1 TC1 = 1.0E-7)
C1   2   3    CMOD   400E-12
.MODEL   CMOD   CAP(C = 1 TC1 = -6.0E-6)
R2   3   0    RMOD   1K
.AC LIN 5000   1.1E6    4E6
.TEMP   25  100
.PRINT AC VM(3)
.PROBE V(3)
.END
```

Partial results for the output voltage at 25°C and 100°C are shown in Table 3.5.
From Table 3.5, the notch frequency at 25°C is 2.51664E + 06 Hz and that at 100°C is 2.51722E + 06 Hz. There is little shift in the notch frequency as the temperature increased from 25°C to 100°C. The magnitude characteristic at 25°C is shown in Figure 3.2.

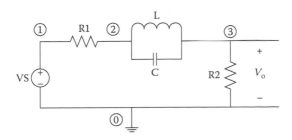

FIGURE 3.1
Notch filter.

TABLE 3.5

Magnitude Characteristics at 25°C and 100°C

Frequency, HZ	Output Voltage at 25°C, V	Output Voltage at 100°C, V
2.51200E+06	2.23325E−02	2.51167E−02
2.51258E+06	1.94193E−02	2.22063E−02
2.51316E+06	1.65048E−02	1.92944E−02
2.51374E+06	1.35893E−02	1.63811E−02
2.51432E+06	1.06730E−02	1.34668E−02
2.51490E+06	7.75638E−03	1.05519E−02
2.51548E+06	4.83962E−03	7.63651E−03
2.51606E+06	1.92304E−03	4.72104E−03
2.51664E+06	9.93068E−04	1.80576E−03
2.51722E+06	3.90840E−03	1.10903E−03
2.51780E+06	6.82266E−03	4.02304E−03
2.51838E+06	9.73556E−03	6.93597E−03
2.51896E+06	1.26468E−02	9.84751E−03
2.52012E+06	1.84631E−02	1.27574E−02
2.52070E+06	2.13676E−02	1.85710E−02

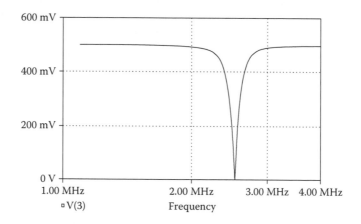

FIGURE 3.2
Notch frequency characteristic at 25°C.

3.1.4 Diode Models

Diode models take into account the forward and reverse bias characteristics of real diodes, the junction capacitance, the ohmic resistance of the diode, the temperature effects on the diode characteristics, and leakage current and high level injection of real diodes. The general format for modeling diodes is

DNAME   NA   NK   MODEL_NAME   [(AREA) VALUE]

.MODEL   MODEL_NAME   D   MODEL_PARAMETERS

where

NA is the node number for the anode;

NK is the node number for the cathode;

MODEL_NAME is a name (preferably starting with D). It can be up to eight characters long;

D is the PSPICE specification for the model associated with diodes;

MODEL_PARAMETERS are parameters that model diode operation. Table 3.6 shows the diode model parameters; and

TABLE 3.6

Diode Model Parameters

Parameters	Description	Default	Unit
IS	Saturation current	1E-14	A
N	Emission current	1	
ISR	Recombination current parameter	0	A
NR	Emission coefficient for ISR	2	
IKF	High-injection "knee" current	Infinite	A
BV	Reverse breakdown "knee" voltage	Infinite	V
IBV	Reverse breakdown "knee" current	1E-10	A
NBV	Reverse breakdown ideality factor	1	
IBVL	Low-level reverse breakdown "knee" current	0	A
NBVL	Low-level reverse breakdown ideality factor	1	
RS	Parasitic resistance	0	Ohm
TT	Transit time	0	s
CJO	Zero-bias p-n capacitance coefficient		F
VJ	p-n Potential	1	V
M	p-n Grading coefficient	0.5	
FC	Forward-bias depletion capacitance coefficient	0.5	
EG	Bandgap voltage (barrier height)	1.11	eV
XTI	IS temperature	3	
TIKF	IKF temperature coefficient (linear)	0	°C^{-1}
TBV1	BV temperature coefficient (linear)	0	°C^{-1}
TBV2	BV temperature coefficient (quadratic)	0	°C^{-2}
TRS1	RS temperature coefficient (linear)	0	°C^{-1}
TRS2	RS temperature coefficient (quadratic)	0	°C^{-2}
KF	Flicker noise coefficient	0	
AF	Flicker noise exponent	1	

AREA_FACTOR is used to determine the number of equivalent diodes that are connected in parallel. Parameters that are affected by the area factor are IS, CJO, IBV, and RS.

For example, the statements:

D1   10   11   DMODEL

.MODEL   DMODEL   D(IS = 1.0E-14   CJO = 3PF
TT = 5NS   BV = 120V   IBV = 5.0E-3)

will model diode D1 with model name DMODEL and model parameters IS, CJO, TT, BV, and IBV. The parameters not supplied will be given the default values.

3.1.5 Bipolar Junction Transistor Models

The PSPICE model of bipolar junction transistors is based on the integral charge-control of Gummel and Poon. For large signal analysis, the Ebers and Moll transistor model can be used. The general format for modeling bipolar transistors is:

QNAME   NC   NB   NE   NS   MODEL_NAME   <AREA_VALUE>

.MODEL   MODEL_NAME   TRANSISTOR_TYPE
MODEL_PARAMETERS

where
NC, NB, NE, and **NS** are the node numbers for the collector, base, emitter, and substrate, respectively;
MODEL_NAME is a name (preferably starting with the character Q). It can also be up to eight characters long;
AREA VALUE is the size of the transistor. It represents the number of transistors paralleled together. The parameters that are affected by the area factor are: IS, IKR, RB, RE, RC, and CJE;
TRANSISTOR_TYPE can be either NPN or PNP; and
MODEL_PARAMETERS are the parameters that model the bipolar junction transistor characteristics.

Table 3.7 shows the bipolar transistor model parameters.
For example, the statements:

Q1   1   3   2   2   QMOD

.MODEL   QMOD   NPN
(IS = 2.0e-14   BF = 20   CJE = 1.5PF   CJC = 200PF   TF = 15NS)

TABLE 3.7

Bipolar Transistor Model Parameters

Parameter	Meaning	Default Value	Unit
IS	p-n Saturation current	1E-16	A
BF	Ideal maximum forward beta	100	
NF	Forward current emission coefficient	1	
VAF(VA)	Forward early voltage	Infinite	V
IKF(IK)	Corner for forward beta high-current roll-off	Infinite	A
ISE(C2)	Base-emitter leakage saturation current	0	A
NE	Base-emitter leakage emission coefficient	1.5	
BR	Ideal maximum reverse beta	1	
NR	Reverse current emission coefficient	1	
VAR(VB)	Reverse early voltage	Infinite	V
IKR	Corner for reverse beta high-current roll-off	Infinite	A
ISC(C4)	Base-collector leakage saturation current	0	A
NC	Base-collector leakage emission coefficient	2.0	
RB	Zero-bias (maximum) base resistance	0	Ohm
RBM	Minimum base resistance	RB	Ohm
IRB	Current at which RB falls halfway to RBM	Infinite	A
RE	Emitter ohmic resistance	0	Ohm
RC	Collector ohmic resistance	0	Ohm
CJE	Base-emitter zero-bias p-n capacitance	0	F
VJE(PE)	Base-emitter built-in potential	0.75	V
MJE(ME)	Base-emitter p-n grading factor	0.33	
CJC	Base-collector zero-bias p-n capacitance	0	F
VJC(PC)	Base-collector built-in potential	0.75	V
MJC(MC)	Base-collector p-n grading factor	0.33	
XCJC	Fraction of Cbc connected internal to RB	1	
CJS(CCS)	Collector-substrate zero-bias p-n capacitance	0	F
VJS(PS)	Collector-substrate built-in potential	0.75	V
MJS(MS)	Collector-substrate p-n grading factor	0	
FC	Forward bias depletion capacitor coefficient	0.5	
TF	Ideal forward transit time	0	s
XTF	Transient time bias dependence coefficient	0	
VTF	Transient time bias dependency on V_{bc}	Infinite	V
ITF	Transit time dependency on I_C	0	A
PTF	Excess phase at $1/(2\pi TF)Hz$	0	°
TR	Ideal reverse transit time	0	s
EG	Bandgap voltage (barrier height)	1.11	eV
XTB	Forward and reverse beta temperature coefficient	0	
XTI(PT)	IS Temperature effect exponent	3	
KF	Flicker noise coefficient	0	
AF	Flicker noise exponent	1	

describes an NPN transistor with model_name QMOD and model parameters specified for IS, BF, CJE, and TF. The undeclared parameters are assumed to have the default values.

The statement:

Q3   3   4   5   5   QMOD2

.MODEL   QMOD2   PNP(IS = 1.0e-15   BF = 50)

describes a PNP transistor with model name QMOD2 and model parameters specified for IS and BF.

3.1.6 MOSFET Models

Depending on the level of complexity, MOSFETS can be modeled by Shichman_Hodges model, geometry-based analytic model, semi-empirical short-channel model, or Berkeley short-channel IGFET model (BSIM). The general format for modeling MOSFETS is:

MNAME   ND   NG   NS   NB   MODEL_NAME
DEVICE_PARAMETERS

.MODEL   MODEL_NAME   TRANSISTOR_TYPE
MODEL_PARAMETERS

where

ND, NG, NS, and **NB** are the node numbers for the drain, gate, source, and substrate, respectively;

MODEL_NAME is a name (preferably starting with the character M). It can be up to eight characters long;

DEVICE_PARAMETERS are optional parameters that can be provided for L (length), W (width), AD (drain diffusion area), AS (source diffusion area), PD (Perimeter of drain diffusion), and RS (Perimeter of source diffusion)

NRD, NRS, NRG, and NRB are the relative resistivities of the drain, source, gate, and substrate in squares. M is the device "multiplier." Its default value is one. It simulates the effect of equivalent MOSFETS connected in parallel;

TRANSISTOR_TYPE can either be NMOS or PMOS; and

MODEL_PARAMETERS selected depends on the MOSFET model that is being used. The LEVEL parameter is used to select the appropriate model. The following models are available:

LEVEL = 1 is for Shichman–Hodges model,

LEVEL = 2 is a geometry-based analytic model,

TABLE 3.8

Model Parameters of MOSFETS

Name	Model Parameter	Unit	Default	Typical
LEVEL	Model type (1, 2, or 3)		1	
L	Channel length	m	DEFL	
W	Channel width	m	DEFW	
LD	Lateral diffusion (length)	m	0	
WD	Lateral diffusion (width)	m	0	
VTO	Zero-biased threshold voltage	V	0	
KP	Transconductance	A/V^2	2E-5	
GAMMA	Bulk threshold parameter	$V^{1/2}$	0	0.35
PHI	Surface potential	V	0.6	0.65
LAMBDA	Channel-length modulation (LEVEL = 1 or 2)	V^{-1}	0	0.02
RD	Drain Ohmic resistance	Ω	0	10
RS	Source Ohmic resistance	Ω	0	10
RG	Gate Ohmic resistance	Ω	0	1
RB	Bulk Ohmic resistance	Ω	0	1
RDS	Drain-source shunt resistance	Ω	∞	
RSH	Drain-source diffusion sheet resistance	Ω/square	0	20
IS	Bulk p-n saturation current	A	1E-14	1E-15
JS	Bulk p-n saturation current/area	A/m^2	0	1E-8
PB	Bulk p-n potential	V	0.8	0.75
CBD	Bulk-drain zero-bias p-n capacitance	F	0	5PF
CBS	Bulk-source zero-bias p-n capacitance	F	0	2PF
CJ	Bulk p-n zero bias bottom capacitance/length	F/m^2	0	
CJSW	Bulk p-n zero-bias perimeter capacitance/length	F/m	0	
MJ	Bulk p-n bottom grading coefficient		0.5	
MJSW	Bulk p-n sidewall grading coefficient		0.33	
FC	Bulk p-n forward-bias capacitance coefficient		0.5	
CGSO	Gate-source overlap capacitance/channel width	F/m	0	
CGDO	Gate-drain overlap capacitance/channel width	F/m	0	
CGBO	Gate-bulk overlap capacitance/channel length	F/m	0	
NSUB	Substrate doping density	$1/cm^3$	0	
NSS	Surface state density	$1/cm^2$	0	
NFS	Fast surface state density	$1/cm^2$	0	
TOX	Oxide thickness	m	∞	
TPG	Gate material type: +1 = opposite of substrate, −1 = same as substrate, and 0 = aluminum		1	

(Continued)

TABLE 3.8

Model Parameters of MOSFETS (Continued)

Name	Model Parameter	Unit	Default	Typical
XJ	Metallurgical junction depth	m	0	
UO	Surface mobility	cm²/V.s	600	
UCRIT	Mobility degradation exponent (LEVEL = 2)	V/cm	1E4	
UEXP	Mobility degradation exponent (LEVEL = 2)		0	
UTRA	(not used) Mobility degradation transverse field coefficient			
VMAX	Maximum drift velocity	m/s	0	
NEFF	Channel charge coefficient		1	
XQC	Fraction of channel charge attributed to drain		1	
DELTA	Width effect on threshold		0	
THETA	Mobility modulation (LEVEL = 3)	V⁻¹	0	
ETA	Static feedback (LEVEL = 3)		0	
KAPPA	Saturation field factor (LEVEL = 3)		0.2	
KF	Flicker noise coefficient		0	1E-26
AF	Flicker noise exponent		1	1.2

LEVEL = 3 is a semi-empirical, short-channel model,

LEVEL = 4 is the BSIM model, and

LEVEL = 5 is the SIM3 model.

Table 3.8 shows the MOSFET model parameters for levels 1, 2, and 3. Discussions on model parameters for LEVEL = 4 and LEVEL = 5 are beyond the scope of this book.

3.2 Library File

Models and subcircuits of devices and components exist in PSPICE. There are more than 5000 device models available that PSPICE users can use for simulation and design. The models exist in different libraries of the PSPICE package. The reader should consult PSPICE manuals for models available for various electronic components.

The .LIB statement is used to reference a model or a subcircuit that exists in another file as a library. The general format for the .LIB command is:

.LIB   FILENAME.LIB

where
   FILENAME.LIB **is the name of the library file.**

If the FILENAME.LIB is left off then the default file is NOM.LIB. The latter
library, depending on the version of PSPICE you are running, will contain
devices or names of individual libraries.
   One can also set up one's own library file using the file extension .LIB. The
device models and subcircuits can be placed in the file. One can access the
individually created libraries the same way the PSPICE supplied libraries are
accessed, i.e., by using:

.LIB FILENAME.LIB.

One should be careful not to give the individually created library file the
same name as the ones supplied by PSPICE. The following example explores
the use of models in a diode circuit.

Example 3.2: Precision Diode Rectifier Characteristics

For the precision diode rectifier, shown in Figure 3.3, VCC = 10 V, VEE = –10 V, X1
and X2 are UA741 op amps. D1 and D2 are D1N4009 diodes. If the input voltage
is a triangular wave, shown in Figure 3.4, with period of 2 ms, peak-to-peak value
of 10 V and zero average value, find the output voltage.

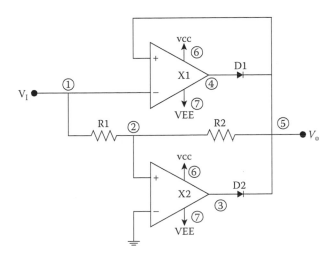

FIGURE 3.3
Precision full-wave rectifier.

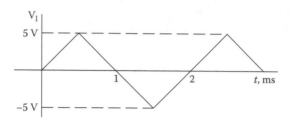

FIGURE 3.4
Input voltage to the full-wave rectifier shown in Figure 3.3.

Solution

PSPICE Program

```
FULL WAVE RECTIFIER
VIN   1   0   PWL(0   0   0.5M   5   1.5M   -5   2.5M   5   3.0M   0)
VCC   6   0   DC  10V
VEE   7   0   DC  -10V
R1    1   2   10K
R2    2   5   10K
X1    1   5   6   7   4   UA741
* + INPUT; -INPUT; + VCC; -VEE; OUTPUT; CONNECTIONS FOR OP
AMP UA741
D1    4   5   D1N4009;   DIODE MODEL IS DIN4009
.MODEL D1N4009 D(IS = 0.1P RS = 4 CJO = 2P TT = 3N BV = 60
IBV = 0.1P)
X2    0   2   6   7   3   UA741
* + INPUT; -INPUT; + VCC; -VEE; OUTPUT; CONNECTIONS FOR OP
AMP UA741
D2    3   5   D1N4009
.TRAN 0.02MS 3MS
.PROBE
.LIB NOM.LIB;
* UA741 OP AMP MODEL IN PSPICE LIBRARY FILE NOM.LIB
.END
```

The PSPICE output is shown in Figure 3.5.

3.3 Component Values (.PARAM, .STEP)

3.3.1 The .PARAM Statement

The .PARAM statement allows one to set component values by using mathematical expression. The general format for the .PARAM statement is:

.PARAM   PARAMETER_VALUE = VALUE

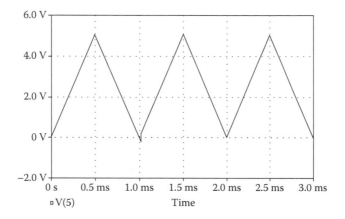

FIGURE 3.5
Output characteristics of precision full-wave rectifier.

or

.PARAM   PARAMETER_NAME = {MATHEMATICAL_EXPRESSION)

where
PARAMETER_NAME is the set of characters allowed by PSPICE; and
PARAMETER_VALUE may be a constant or a mathematical expression.
The intrinsic functions that can be used to form mathematical expressions are shown in Table 3.9.

For example, the statement:

.PARAM   C1 = 1.0UF,   VCC = 10V, VSS = –10V

defines the values of C1 as 1.0 µF, VCC = 10 V, and VSS = –10 V.
For the following two statements:

.PARAM RA = 10K

.PARAM RB = {5*RA}

the first statement sets RA to be 10 Kilohms and the second statement equates RB to be 5 times RA.
The following points should be observed while using the parameter definition statement, PARAM:

1. If the parameter is defined by an expression, then the curly brackets { } are required.
2. .PARAM can be used inside a subcircuit definition to create parameters that are local to the subcircuit.

TABLE 3.9

Valid Functions for Mathematical Expression

Function	Meaning	Comments
ABS(X)	$\lvert X \rvert$	Absolute value of X
ACOS(X)	$\cos^{-1}(X)$	X is between −1 and 1
ARCTAN(X)	$\tan^{-1}(X)$	Results in radians
ASIN(X)	$\sin^{-1}(X)$	$-1.0 \leq X \leq 1.0$
ATAN(X)	$\tan^{-1}(X)$	Results in radians
ATAN2(Y,X)	$\tan^{-1}(Y/X)$	Results in radians
COS(X)	$\cos(X)$	X in radians
DDT(X)	Time derivative of X	
EXP(X)	e^X	
IMG(X)	Imaginary part of X	Returns 0.0 for real number
LOG(X)	$\ln(X)$	log base e of X
LOG10(X)	$\log(X)$	log base 10 of X
MAX(X,Y)	Maximum of X and Y	
MIN(X,Y)	minimum of X and Y	
M(X)	$\lvert X \rvert$	Magnitude of X, same as ABS(X)
P(X)	phase of X	Returns 0.00 for real numbers
PWR(X,Y)	$\lvert X \rvert^Y$	Absolute value of X raised to the power of Y
R(X)	Real part of X	
SDX(X)	Time integral od X	Only used for transient analysis
SGN(X)	Signum function	
SIN(X)	$\sin(X)$	X is in radians
SQRT(X)	$X^{1/2}$	Square root of X
TAN(X)	$\tan(X)$	X is in radians

3. There are PSPICE predefined parameters, such as TEMP, VT, GMIN, TIME. Parameter names defined with .PARAM statements should be different in name from the PSPICE predefined parameters.

4. Once defined, a parameter can be used in place of all numeric values in a circuit description. For example:

.PARAM   TWO_PI = {2.0*3.14159}, F0 = 5KHZ

.PARAM   FREQ = {TWO_PI *FO}.

5. .PARAM statements can be in a library. If the PSPICE simulator does not find parameters in a circuit file, it will search libraries for the parameters.

3.3.2 .STEP Function

The .STEP function can be used to vary the circuit element, signal source or a temperature over a specified range. This feature of PSPICE allows the user to observe effects of changing a circuit element on the response of the circuit. The general format for using the statement is:

.STEP   SWEEP_TYPE   SWEEP_NAME

START_VALUE   END_VALUE   INCNP

or

.STEP SWEEP_NAME LIST   < VALUES >

where
SWEEP_TYPE **can be LIN, OCT, DEC;**
For **LIN**, we have linear sweep. The sweep variable is swept linearly with the **INCNP** being the step size from the starting value to ending value;
For **OCT**, (SWEEP by OCTAVE), the sweep variable is swept logarithmically by octave from the start value to ending value. However, **INCNP** is now the number of steps per octave;
For **DEC**, (SWEEP by DECADE), the sweep variable is swept logarithmically by decade from the start value to ending value. **INCNP** is number of steps per decade; and
SWEEP_NAME is the sweep variable name. It can be a model parameter, temperature, global parameter and independent voltage, or current source. During the sweep, the source's voltage or current is set to the sweep value. For example, the statements:

VCE   5   0   DC   10V

.STEP VCC 0   10   2

will cause VCC to be swept linearly from 0 to 10 V with 2 V steps.

MODEL_PARAMETER—Model type and model name are followed by a model parameter name in parenthesis. The parameter in the model is set to the sweep value. For example, the statements:

R1   5   6   RMOD   1

.MODEL   RMOD   RES(R = 1)

.STEP   RES   RMOD(R)   1000   3000   500

will cause PSPICE to analyze circuits for the following values of R1: 100, 1500, 2000, 2500, and 3000 Ohms. In the above example, RMOD is the model_name.

RES is model type and R is the parameter within the model to step. It should be noted that the:

$$\text{Computed value of } R1 = \text{Line value of } R1 \text{ multiplied by } R \qquad (3.8)$$

where
   Line value of $R1$ is the value at the end of the statement line that describes R1; and
   R is the value of R in the model statement.

In another example, the statements:

C2   2   0   CMODEL   1.0e-9

.MODEL   CMODEL   CAP   (C = 1)

.STEP   CAP   CMODEL(C)   2   22   4

will cause PSPICE to analyze the circuits with the above statements for the following values of C2: 2e-9, 6e-9, 10e-9... to 22e-9 Farads.
   TEMPERATURE - Keyword **TEMP** is used for the sweep variable name. It should be followed by the keyword **LIST**. The temperature is set to the sweep value. For each value in the sweep, all the circuit components have their model parameters updated to that temperature. For example, the statements:

V1   1   0   DC   5

R1   1   2   1K

C1   2   0   1U

.STEP   TEMP   LIST   0   27   50   100

.OP

imply that for each temperature value 0, 27, 50, and 100, the component values will be updated and the operating point of the component will be determined.
   GLOBAL_PARAMETER - Keyword PARAM followed by the parameter name. The latter parameter name is set to sweep. During the sweep, the global parameter's value is set to the sweep value and all expressions are evaluated. For example, the statements:

VIN   1   0   AC   1

R1   1   2   100

.PARAM   (CVAL = 5e-6; original value of C1)

C1   2   0   {CVAL}

.STEP   PARAM   CVAL   4e-6   16e-6   2e-6

*   Vary C1 from 4e-6 to 16e-6 by steps of 2e-6F

.AC   LIN   100   1e4   5e4

.PROBE   V(2)

.END

will obtain the value of the voltage V(2) for values of C1 swept from 4 μF to 16 μF with 2 μF steps.

The following points should be noted while using the .STEP statement.

1. The .STEP statement causes all analyses specified in a circuit file (.DC, .AC, .TRAN) to be done for each step.
2. The start value may be less than or greater than the end value of the .STEP statement.
3. The sweep increment value or the number of points per decade or octave (INCNP) should be greater than zero.

Example 3.3: Effect of Damping on RLC Circuit

For the RLC circuit shown in Figure 3.6, R = 1 Ohm, L = 1 Henry, and the initial voltage across capacitor is 3.3 V. If the capacitor C assumes the values of 1, 2, and 3 Farads, determine the voltage across the capacitor with respect to time.

Solution
PSPICE Program

```
THREE CASES OF DAMPING
**
L   1   0   1
R   1   2   1
C   2   0   {C1}   IC = 3.3V
.PARAM   C1 = 1.0; ORIGINAL VALUE
.STEP   PARAM   C1  1  3   1;   VARY C1 FROM 1,2,3 F
.TRAN   0.1 10 UIC
.PLOT   TRAN   V(2)
.PROBE   V(2)
.END
```

The plot of the three cases of damping for RLC circuit is shown in Figure 3.7.

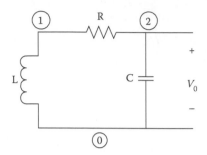

FIGURE 3.6
RLC circuit.

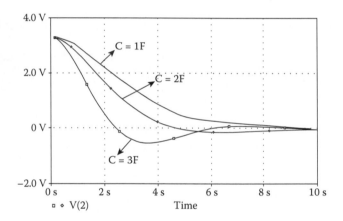

FIGURE 3.7
Transient response of an RLC circuit.

3.4 Function Definition (.FUNC, .INC)

3.4.1 .FUNC Statement

The function statement is used to define "functions" that may be used in expressions similar to those discussed in Section 3.3.1. The functions are user defined and are flexible. Since expressions are restricted to a single line, several sub-expressions can be defined in a circuit file using the function statement to obtain an expression that satisfy ones application. The general form of the function statement is:

.FUNC   FUNC_NAME(ARG)   {BODY}

where
>**FUNCT_NAME** is a name of the function with argument, ARG. The FUNCT_NAME must be different from built-in functions, shown in Table 3.9;
>
>**ARG** is the argument for the function. Up to 10 arguments may be used in a definition. The number of arguments in a function must agree with the number in the function definition. A function may be defined with no arguments, but the parenthesis is still required; and
>
>**BODY** of a function definition may refer to other functions previously defined. The body of a function is enclosed in curly braces { }.

The following points should be borne in mind using the .FUNC statement.

1. The .FUNC statement must precede the first use of the function name **FUNCT_NAME**.

2. The body of a function definition must fit on one line.

3. FUNC statements if they appear in subcircuits, are local to those subcircuits.

If an application has several .FUNC statements, the user can create a file that contains the .FUNC definitions and access the function definitions with a .INC statement near the beginning of the circuit file. The next section describes the .INC statement. The following example describes the use of the .FUNC statement.

Example 3.4: Thermister Characteristics

Thermister is a device whose resistance is highly dependent on temperature. It can be used to measure temperature. The resistance, R_T, of a thermister can be expressed as:

$$R_T = R_O \exp\left(\beta\left(\frac{1}{T} - \frac{1}{T_O}\right)\right) \tag{3.9}$$

where
>R_T is the resistance of a thermister at temperature T degrees Kelvin;
>R_O is the resistance at T_O in degrees Kelvin, which is usually taken to be 298°K ($\equiv$ 25°C); and
>β is a characteristic temperature that varies with material composition of a thermister. A typical value is 4000°K, but it can vary from 1500°K to 6000°K.

Figure 3.8 shows a simple thermister circuit. VS = 10 V and RS = 25 KΩ. The thermister resistance at temperature 298°K is 25 KΩ. The characteristic temperature of

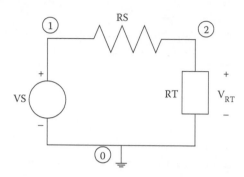

FIGURE 3.8
Thermister circuit.

the thermister is 4000°K. Determine the voltage across the thermister with respect to temperature.

Solution

The PSPICE program for obtaining the thermister characteristics is as follows:

PSPICE Program

```
THERMISTER CIRCUIT -
*     TO = 298
*     RO = 25000
*     B = 4000
.PARAM  TS = 300
* Calculates resistance of thermister at temperature TS
.FUNC E(X) {EXP(X)}
.FUNC RT(Y) {25000*(E(4000*((1/Y) - 0.00336)))}
.STEP PARAM TS 300 400 10
*
VS    1    0    DC    10
R1    1    2    25K
RT    2    0    {RT(TS)}
.DC   VS   10   10    1
.PRINT DC V(2)
.END
```

Table 3.10 shows the temperature versus output voltage of the thermister circuit.

3.4.2 .INC Statement

The .INC statement may be used to insert the content of another file into a circuit file. The general format for using the .INC statement is:

.INC "FILENAME"

TABLE 3.10

Voltage versus Temperature of a Thermister

Temperature, °K	Voltage Across Thermister, V
300	4.734
310	3.689
320	2.809
330	2.110
340	1.577
350	1.180
360	0.888
370	0.673
380	0.515
390	0.398
400	0.311

where

FILENAME is a character string that is a legal filename for the computer system the user is running the PSPICE package on.

The included file may contain PSPICE valid statements with the following exceptions:

(a) No title line is allowed. Instead of the title line, a comment line may be used.

(b) .END statement is not required. However, if an .END statement is present it should mark the end of the included file.

(c) .INC statement may be used in the included file. However, only up to four levels of "including" are allowed.

It should be noted that including a file by using the .INC statement brings the file's text into the circuit file and takes up space in main memory (RAM).

3.5 Subcircuit (.SUBCKT, .ENDS)

If a block of circuit is repetitively used in an overall circuit, then the block circuit can be defined as a subcircuit. The subcircuit can then be used repetitively. The subcircuit concept is similar to those of subroutines in programming languages, such as Fortran or C. The general form for subcircuit description is:

.SUBCKT   SUBCIRCUIT_NAME   NODE1   NODE2 ...
[PARAMS:NAME – <VALUE>]

DEVICE   STATEMENTS

.ENDS   [SUBCIRCUIT_NAME].

The subcircuit definition begins with **.SUBCKT** statement. The subcircuit definition ends with **.ENDS** statement. The statements between .SUBCKT and .ENDS are included in the subcircuit definition.

Subcircuit definitions contain only device statements. Subcircuit definitions may also contain .MODEL, .PARAM, or .FUNC statements.

Ends subcircuit name indicates the end of subcircuit circuit description statements. The subcircuit_name after .END statement may be omitted. However, it is advisable to have the subcircuit definition that is being terminated. This is especially useful if there are more than one subcircuit that is called by the main circuit.

The symbol for a subcircuit call is **X**. It is essential that a unique name be given to each separate call of a subcircuit. The general form for a subcircuit call is:

XNAME   NODE1   NODE 2   ..   SUBCIRCUIT_NAME
[PARAMS:NAME = VALUE]

where
XNAME is the device name of the subcircuit. Xname can be up to eight characters long;

NODE1, NODE2. There must be the same number of nodes in the subcircuit calling statement as in its definition. The nodes in the subcircuit call must match those of the subcircuit definition; and

SUBCIRCUIT_NAME is the name of the subcircuit to be inserted into the main circuit (or calling circuit).

It should be noted that subcircuits may be nested. That is subcircuit X may call other subcircuits. In addition, nesting of subcircuits cannot be circular. That is if subcircuit X contains a call to subcircuit Y, then subcircuit Y must not contain a call to subcircuit X.

Subcircuits have the following advantages:

1. It reduces the size of circuit file, provided the circuit has repetitive elements. The repetitive parts of the circuit can be defined as a subcircuit and the main circuit issue subcircuit calls.

2. Once a subcircuit is defined, it can be used by other circuits. For example, once a subcircuit of a particular op amp is known, it can be used to build and simulate amplifiers, oscillators, and filters that use that particular op amp.

3. Subcircuits allow hierarchal testing and design of complex circuits. A complex circuit can be divided into the sum of parts. If some of the parts are repetitive, subcircuit definitions can be done for those

repetitive parts. Once the various parts of the complex circuit are simulated and their functionality validated, then they can be used to build more complex circuits.

The following example illustrates the use of subcircuits.

Example 3.5: Frequency Response of a State-Variable Active Filter

For the state-variable active filter shown in Figure 3.9, the op amp has an input impedance of 10^{12} Ohms, an open loop gain of 10^7 and a zero output resistance. $R1 = 80$ KΩ, $R2 = 15$ KΩ, $R3 = 500$ Ω, $R4 = 5$ KΩ, $R5 = 200$ KΩ, $R6 = 5$ KΩ, $R7 = 200$ KΩ, and $C1 = C2 = 2$ nF. Determine the frequency response of the filter.

Solution

The op amp symbol and its simplified equivalent circuit is shown in Figure 3.10. The subcircuit definition is:

```
.SUBCKT  OPAMP  1  2  3
* - INPUT; + INPUT; OUTPUT
RIN  1  2  1.0E12
EVO  0  3  1  2  1.0E7
.ENDS  OPAMP
```

PSPICE Program

```
STATE-VARIABLE
VIN   1   0    AC  1   0
R1    1   2    80K
X1    2   3    4    OPAMP
R2    2   4    15K
R3    3   0    500
R4    4   5    5K
X2    5   0    6    OPAMP
C1    5   6    2nF
R5    3   6    200K
R6    6   7    5K
X3    7   0    8    OPAMP
C2    7   8    2nF
R7    8   2    200K
.AC    DEC 10  1E2 1E6
.PRINT     AC VM(6)
.PROBE
.SUBCKT OPAMP   1   2   3
* - INPUT; + INPUT; OUTPUT
RIN   1   2    1E12
EVO   0   3    1   2   1.0E7
.ENDS OPAMP
.END
```

The magnitude response of the filter is shown in Figure 3.11.

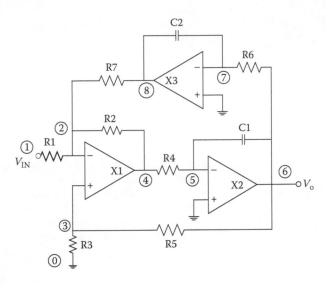

FIGURE 3.9
State-variable active filter.

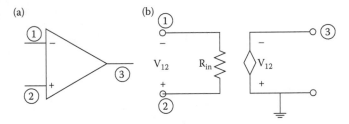

FIGURE 3.10
Op amp (a) block diagram, and (b) its simplified equivalent circuit.

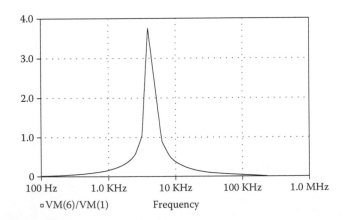

FIGURE 3.11
Frequency response of a state-variable filter.

3.6 Analog Behavioral Model

One way of simulating circuits is to describe the circuit in terms of the components and the connection between the components. The latter can be resistors, capacitors, inductors, transistors, voltage, and current sources. This type of simulation is called **structure or primitive-level simulation**. Most of the PSPICE simulations, performed so far, were the type of primitive-level simulation. If the circuit contains a lot of elements, the primitive-level simulation takes a long simulation time.

When a designer is interested in a system performance, primitive-level simulation of all the subsystems may be too detailed, time-consuming, and ineffective. At the system level, a block-diagram simulation approach might be desirable and appropriate. The function of the blocks representing a system may be described by their mathematical behavior, expressions, or relations. This type of simulation is called **analog behavioral modeling simulation**. This section discusses the analog behavioral model simulation approach of PSPICE.

The general format for using the analog behavioral model (ABM) is:

ENAME   CONNECTING_NODES

ABM_KEYWORD   ABM_FUNCTION

or

GNAME   CONNECTING_NODES

ABM_KEYWORD   ABM_FUNCTION

where
>   ENAME **or** GNAME **is the component name assigned to the E or G device;**
>   **CONNECTING_NODES** specify the " + node" and "-node" between which the component is connected;
>   **ABM_KEYWORD** specifies the form of transfer function to be used. One of the following functions may be used,

- VALUE: Arithmetic expression
- TABLE: Lookup table
- FREQ: Frequency response
- LAPLACE: Laplace transform; and

>   **ABM_FUNCTION** specifies the transfer function as a mathematical expression, lookup table, or ratio of two polynomials.

The two controlled sources that may be used for the ABM are voltage-controlled voltage source or voltage controlled current source. Thus for modeling voltage sources, the PSPICE component should start with the letter **E**. Similarly, for modeling current sources, the PSPICE component should start with the letter **G**. The following subsection will describe the analog behavioral model functions.

3.6.1 Value Extension

The VALUE extension allows transfer function to be written as mathematical expressions. The general forms are:

$$\text{ENAME} \quad \text{N} + \text{N-} \quad \text{VALUE} = \{(\text{EXPRESSION})\}$$

or

$$\text{GNAME} \quad \text{N} + \text{N-} \quad \text{VALUE} = \{(\text{EXPRESSION})\}$$

where
(EXPRESSION) is a mathematical expression that may contain arithmetic operators (+ ,-,*,/), .PSPICE built-in function shown in Table 3.9, constants, node voltages, currents, and the parameter TIME. The latter variable is PSPICE interval sweep variable that is employed in transient analysis.

It should be noted that:

(a) VALUE in the statement line should be followed by a space.
(b) (expression) must fit on a single line. If it cannot, start the following line by + and continue writing the value of the expression.

Some valid VALUE statements are:

$$\text{EAVE} \quad 1 \quad 0 \quad \text{VALUE} = \{.25*V(2,0) + (V(2,0) + V(2,3) + V(3,0)\}$$

$$\text{GVMW} \quad 4 \quad 0 \quad \text{VALUE} = \{10*\cos(6.28*\text{TIME})\}.$$

The following example illustrates the use of the VALUE extension.

Example 3.6: Voltage Multiplier

A voltage multiplier has an output given as

$$V_0 = k[V_1(t) * V_2(t)] \tag{3.10}$$

If $V_1(t)$ and $V_2(t)$ are triangular and sinusoidal waveforms, find the output voltage. Assume that $k = 0.4$ and $R_0 = 100$ Ohms.

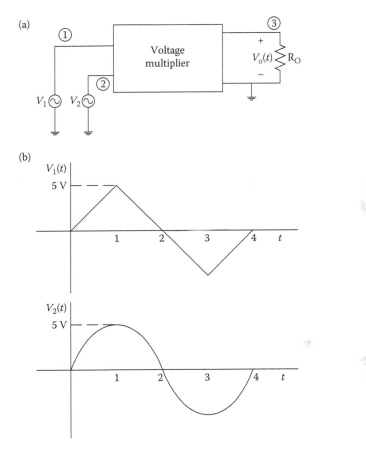

FIGURE 3.12
(a) Multiplier block diagram, (b) input signals $V_1(t)$ and $V_2(t)$.

Solution

PSPICE Program

```
VOLTAGE MULTIPLIER
V1   1   0   PWL(0   0   1MS   5V   3MS   -5V   5MS   5V   6MS   0)
.PARAM K = 0.4
V2   2   0   SIN(0   5   250   0   0   0)
* MULTIPLIER MODEL
EMULPLY 3   0   VALUE = {K*V(1,0)*V(2,0)}
R0   3   0   100
.TRAN   0.02MS 6MS; TRANSIENT RESPONSE
.PROBE
.END
```

The output voltage is shown in Figure 3.13.

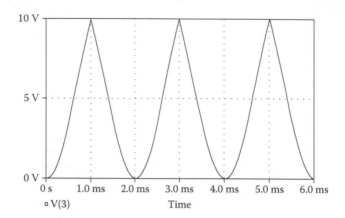

FIGURE 3.13
Output of a voltage multiplier.

3.6.2 Table Extension

The TABLE extension can be used to describe the operation of a circuit or a device by a lookup table. The general form of the table function is:

ENAME   N+ N−   TABLE{EXPRESSION}
= <<INPUT   VALUE> <OUTPUT   VALUE>>

or

GNAME   N+ N−   TABLE{EXPRESSION}
= <<INPUT   VALUE> <OUTPUT   VALUE>>

where
   N +, N- are the positive and negative nodes between which the compo-
      nent is connected;
   TABLE is the keyword showing that the controlled sources are described
      by a tabular data; and
   EXPRESSION is the input to the table, which is evaluated based in the
      tabular data. The table itself consists of pairs of values. The first value
      in each pair is the input and the second is the corresponding output.

It should be noted that:

1. The table's input must be in order from lowest to highest.
2. Linear interpolation is performed between entries.
3. The TABLE extension can be used to describe circuits or devices that
   can be represented by measured data.

4. TABLE keyword must be followed by a space.

5. The input to the table is < expression > that must fit on one line.

The following example uses the TABLE expression to find a diode current.

Example 3.7: Current in Diode Circuit

In Figure 3.14, R1 = 5 KΩ, R2 = 5 KΩ, R3 = 10 KΩ, and R4 = R5 = 10 KΩ. Table 3.11 describes the current–voltage characteristics of the diode. Find the current flowing through the diode.

Solution

PSPICE Program

```
DIODE CIRCUIT
V1   1   0    DC  15V
R1   1   2    5K
R2   2   0    5K
R3   2   3    10K
R4   3   0    10K
GDIODE  3   4    TABLE {V(3,4)} = (0 0) (0.1 0.13E-11) (0.2 1.8E-11)
+ (0.3  24.1E-11)  (0.4 0.31E-8)  (0.5  4.31E-8)  (.6 58.7E-8)
+ (0.7 7.8E-6)
R5   4   0    10K
.DC V1 15  15  1
.PRINT  DC I(R5)
.END
```

Relevant results from PSPICE simulation are:

```
V1 I(R5)
1.500E + 01 7.800E-06
```

thus, the current flowing through the diode is a 7.8E-06 A.

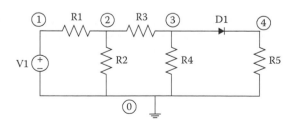

FIGURE 3.14
A diode circuit.

TABLE 3.11

Diode Characteristics

Forward Voltage in Volts	Forward Current in Amps
0	0
0.1	0.13e–11
0.2	1.8e–11
0.3	24.1e–11
0.4	0.31e–8
0.5	4.31e–8
0.6	58.7e–8
0.7	7.8e–6

3.6.3 FREQ Extension

The FREQ function can be used to describe the operation of a circuit or system by a frequency response table. The general form of the FREQ function is:

> **ENAME   N+ N–   FREQ   {EXPRESSION}**
> **= FREQUENCY VALUE, MAGNITUDE IN DB, PHASE   VALUE**

or

> **GNAME   N+ N–   FREQ   {EXPRESSION}**
> **= FREQUENCY VALUE, MAGNITUDE IN DB, PHASE VALUE**

where

   N+, N– are the positive and negative nodes between which the component is connected;

   FREQ is the keyword showing that the controlled sources are described by a table of frequency response; and

   EXPRESSION is the input to the table. The table consists of frequency value and its corresponding magnitude in decibels(dB) and phase (degrees). Interpolation is done between entries. Phase is linearly interpolated and the magnitude logarithmic interpolated. The frequencies in the table must be in order from lowest to highest.

The FREQ and TABLE functions are similar in use. Both functions are described by tabular data. FREQ function is used to describe a circuit or system in terms of frequency response points (frequency, magnitude, phase). However, the TABLE function is used to describe circuit, device, or system operation in terms of (x, y) values.

Note that:

1. FREQ keyword must be followed by a space.

2. EXPRESSION must fit on one line.

The following example describes the application of FREQ function to plot an experimental data obtained from a filter.

Example 3.8: Frequency Response of a Filter

A block diagram of a filter is shown in Figure 3.15. Data obtained from the filter are shown in Table 3.12. Plot the magnitude and phase responses.

Solution

PSPICE Program

```
FILTER CHARACTERISTICS
VIN       1    0    AC    1    0
R1    1    0    1K
EFILTER 2    0    FREQ {V(1,0)} = (1.0K, -14, 107) (1.9K, -9.6, 90)
+ (2.5K, -5.9, 72) (4.0K, -3.3, 55) (6.3K, -1.6, 39) (10K,
-0.7, 26)
 + (15.8K, -0.3, 17) (25K, -0.1, 11) (40K, -0.05, 7) (63K,
-0.02, 4) (100K, -0.008, 3)
R2    2    0    1K
*INPUT NODES 1 AND 0, AND OUTPUT IS BETWEEN NODES 2 AND 0.
.AC   DEC   5   1000   1.0E5
.PROBE V(2) V(1)
.END
```

The magnitude and phase responses are shown in Figure 3.16a and b.

3.6.4 LAPLACE Extension

The LAPLACE function can be used to describe the operation of a circuit or system by means of a transfer function given in terms of Laplace transform function. The general form of the LAPLACE function is:

ENAME   N+   N– LAPLACE{EXPRESSION} = {TRANSFORM}

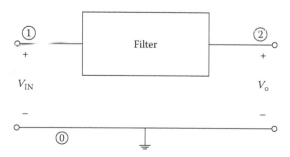

FIGURE 3.15
Block diagram of a filter.

TABLE 3.12

Frequency Response of a Filter

Frequency, Hz	Magnitude, dB	Phase in Degrees
1.0 K	−14	107
1.9 K	−9.6	90
2.5 K	−5.9	72
4.0 K	−3.3	55
6.3 K	−1.6	39
10 K	−0.7	26
15.8 K	−0.3	17
25 K	−0.1	11
40 K	−0.05	7
63 K	−0.02	4
100 K	−0.008	3

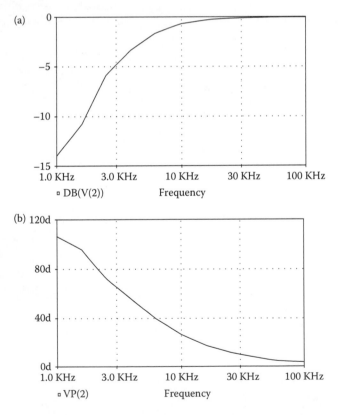

FIGURE 3.16

(a) Magnitude response, (b) phase response of a filter.

or

GNAME N+ N– LAPLACE {EXPRESSION} = {TRANSFORM}

where
  N + N– are positive and negative nodes between which the component is connected;
  LAPLACE is the keyword showing that the controlled sources are described by Laplace transfer variable S;
  EXPRESSION is the input to the transform. It follows the same rules mentioned in Section 3.6.1. It can be voltage, current, or mathematical expression containing voltage, current arithmetic operators, and PSPICE built-in functions; and
  TRANSFORM is an expression given by a ratio of two polynomials in the Laplace variable S.

Both LAPLACE and FREQ functions can be used to model frequency response of a circuit or system. LAPLACE function is appropriate if the transfer function of the circuit or system is given in terms of Laplace transform variable S. On the other hand, the FREQ function is used if the transfer function of the circuit or system is given in terms of frequency response table. The LAPLACE function lends itself to changing the polynomial coefficients to ascertain their effects on the circuit response. The following example describes the use of the LAPLACE function.
  Note that:

1. LAPLACE must be followed by a space.
2. EXPRESSION and TRANSFORM must each fit on one line.
3. Voltage, current, and TIME must not be used in a LAPLACE transfer.
4. One can use both the transient and AC analysis with the LAPLACE function.

Example 3.9: Laplace Transform Description of a Bandpass Filter

The voltage transfer function of a second-order band pass filter is (see Figure 3.17):

$$\frac{V_{OUT}}{V_{IN}} = \frac{As}{s^2 + Bs + C} \tag{3.11}$$

where
  s is the Laplace transform variable and A, B, and C are expressions describing the filter characteristics.

If $A = (R/L)$, $B = (R/L)$, $C = (1/LC)$, $L = 5$ H, $R = 100$ Ω, and $C = 10$ μF; plot the magnitude response.

Solution

PSPICE Program

```
FREQUENCY RESPONSE OF A FILTER
VIN   1     0    AC   1
.AC   DEC   20   1     10K;
*FILTER CONSTANTS
.PARAM A = {100/5}
.PARAM B = {100/5}
.PARAM C = {1/5.0E-6}
*FILTER TRANSFER FUNCTION
EBANPAS 2 0    LAPLACE {V(1,0)} = {A*S/(S*S + S*B + C)}
.PROBE V(2)    V(1)   ; FILTER OUTPUT
.END
```

The magnitude response is shown in Figure 3.18.

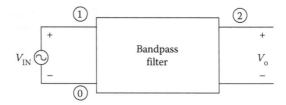

FIGURE 3.17
Laplace transform description of a filter.

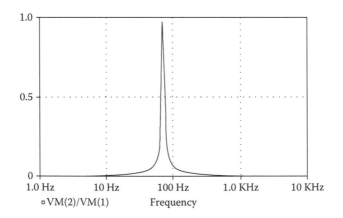

FIGURE 3.18
Frequency response of a bandpass filter.

3.7 Monte Carlo Analysis (.MS)

The parameters of electric element and electronic devices vary due to tolerances incurred from manufacturing processes and also due to aging of components. The Monte Carlo analysis allows the user to vary device parameters and to observe the overall system for variations in circuit parameters. The general form of the Monte Carlo analysis statement is:

> .MC   NUM_RUNS   ANALYSIS   OUTPUT_VARIABLE
>
> FUNCTION   OPTIONS   [SEED = VALUE]

where

.MC statement causes Monte Carlo (statistical) analysis of a circuit to be done;

NUM_RUNS is the number of runs of the selected analysis (DC, AC, or transient). The first run is done with nominal values of all components. Subsequent runs are done with variations of DEV and LOT tolerances on each .MODEL parameter. For printed results, upper limit of NUM_RUNS is 2000. However, for PROBE results, the limit is 400;

ANALYSIS is the analysis type that must be one of these: DC, AC, or TRAN. The specified analysis type is repeated in the *subsequent* passes of the analysis;

OUTPUT_VARIABLE is the output variable that is to be tested. It is identical in format to that of a .PRINT output variable, discussed in Section 2.5; and

FUNCTION specifies the operation to be performed on the OUTPUT_VARIABLE in order to reduce the values of the latter to a single value. The function must be one of the following.

(I)   YMAX - is the greatest difference in each waveform from the nominal value.

(II)   MAX - is the maximum value of each waveform.

(III)   MIN - is the minimum value of each waveform.

(IV)   RISE_EDGE < value > - is the first occurrence of the waveform above a threshold value.

(V)   FALL_EDGE < value > - is the first occurrence of the waveform below a threshold value.

It should be noted that FUNCTION has no effect on the PROBE data saved from the simulation;

OPTIONS are additional items that can be requested during the Monte Carlo analysis. Options include none or more of the following:

(I)   LIST - will print out, at the beginning of each run, the model parameter values used during each run.

(II) OUTPUT (output_type) - request output from runs subsequent to the nominal (first) run. Output_type can be one of the following:

   (a)   ALL - all outputs for all runs are generated including the nominal runs.

   (b)   FIRST < value > - generates outputs only for the first runs where n is specified by < value >,

   (c)   EVERY < value > - generates output every nth value, where n is specified by < value >.

   (d)   RUNS < value > - does analysis and generate output only for specified runs, given by < value >. Up to 25 values may be specified in a list.

   (e)   RANGE(< low_value >, < high value >)- restricts the range over which < FUNCTION > will be evaluated. An "*" can be used in place of either < low_value > or < high-value > to indicate for all values. If RANGE is omitted, then < FUNCTION > is evaluated over the whole sweep range. This is equivalent to RANGE(*,*); and

[SEED = VALUE] - is the seed value for the random number generator within the Monte Carlo analysis. The default value is 17,533. For almost all analysis, it is advisable to use the default seed value in order to achieve a constant set of results.

3.7.1 Component Tolerances for Monte Carlo Analysis

In Monte Carlo analysis, device parameters are allowed to change. There are two ways of changing device parameters in PSPICE: (1) deviation of device parameters **(DEV)**, and (ii) deviation of lot parameters **(LOT)**. The **DEV** parameter allows components to vary independently of other components. The **LOT** parameter allows components from the same lot to track each other.

.MODEL statement of PSPICE is used to assign tolerances of components. The model statement was discussed in Section 3.1. The general format of including the DEV and LOT parameters in the .MODEL statement is:

[DEV/DISTRIBUTION] < VALUE > [%]
[LOT/DISTRIBUTION] < VALUE > [%]]

where

DISTRIBUTION parameters can be one of the following:

UNIFORM - generates uniform distributed deviations over the range $\pm <$ value $>$.

GAUSS - generates deviations with a Gaussian distribution over the range $\pm 3\sigma$ and $<$ value $>$ specifies the $\pm 1\sigma$ deviation. PSPICE limits the values of Gaussian distributions to $\pm 4\sigma$, where σ is the standard deviation.

USER_DEFINED DISTRIBUTION - generates deviations using a user_defined distribution and $<$ value $>$ specifies the ± 1 deviation in the user_defined_distribution; and

VALUE parameter is the deviation of the component value in percentages.

The following examples illustrate how device parameters can be changed in PSPICE.

(1) *DEV tolerance used with .MODEL statement*

R1   1   2   RMOD1   10K

R2   2   3   RMOD1   50K

.MODEL   RMOD1   RES(R = 1   DEV = 10%)

During statistical analysis runs, the value of R1 and R2 are varied at most by 10%. The variations in R1 and R2 will be independent. In the simulation, R1 can take any value from 9K and 11K and R2 any values from 45K to 55K.

(2) *LOT tolerance used with .MODEL statement*

R3   3   4   RMOD1   15K

R4   4   5   RMOD!   20K

.MODEL   RMOD1   RES(R = 1   LOT = 5%)

During statistical analysis runs, the value of R3 and R4 are varied by at most 5%. However, R3 and R4 will increase or decrease by the same percentage.

(3) *DEV and LOT tolerances used with .MODEL statement*

C1   10   11   CMOD   50nF

C2   11   12   CMOD   100nF

.MODEL   CMOD   CAP(C = 1 LOT = 1% DEV = 5%)

During simulations, C1 and C2 are assigned LOT variations up to 1% and DEV variations up to 5%. The two tolerances add. Thus, C1 and C2 can be up to 6% from their nominal values. The 5% tolerance in DEV makes the changes in C1 and C2 uncorrelated, but the 1% tolerance in LOT C1 and 2 makes the change in C1 and C2 correlated.

3.7.2 Simulation

There is a trade-off between the qualities of statistical data obtained from the Monte Carlo analysis versus the simulation time. In order to obtain realistic estimations of the true maximum and minimum limits of component or system variation, it is necessary to perform several runs of the Monte Carlo analysis. However, it takes longer to do more runs. The simulation time is proportional to the number of runs to be performed. The user has to make intelligent decisions as to the number of runs and the time needed for a simulation.

Three types of data are available from a Monte Carlo analysis.

(1) The .OUT file contains model parameters with tolerances applied.
(2) Using PROBE, .PRINT, and .PLOT, the waveforms for each run are available for viewing, printing, or plotting.
(3) The .OUT file may contain the summary of all the runs and may be obtained using Monte Carlo function statements, described in section 3.7.

The following example shows the application of the Monte Carlo analysis.

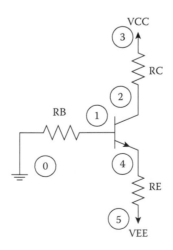

FIGURE 3.19
Universal biasing network.

Example 3.10: Monte Carlo Analysis of a Bipolar Transistor Biasing Network

A universal bipolar biasing network is shown in Figure 3.19. RB = 10 K, RE = 1 K, RC = 1 K, VCC = 10 V, and VEE = −10 V. The resistors have 5% tolerance with uniform distribution. If the beta of the transistor, β_{fr} is 100 and device variation is 10% with uniform distribution, find the changes in the biasing point.

Solution

PSPICE Program

```
MONTE CARLO ANALYSIS
*CIRCUIT ELEMENTS
VCC   3   0    DC 10V
VEE   5   0    DC -10V
RB    1   0    RMOD    10K
RC    3   2    RMOD    1K
RE    4   5    RMOD    1K
Q1    2   1    4    QMOD
*MODEL OF DEVICE WITH TOLERANCES
.MODEL RMOD RES(R = 1DEV/UNIFORM 5%)
.MODEL QMOD   NPN (BF = 100 DEV/UNIFORM  10%VJC = 0.7V)
*MONTE CARLO ANALYSIS
.MC    100    DC    I(RC)    MAX    LIST    OUTPUT    ALL
*100 RUNS, MONITOR CURRENT THROUGH RC USING YMAX COLLATING
FUNCTION
.DC VCC 10    10    1
.PRINT   DC V(2, 4)    I(RC)
.END
```

The edited version of the PSPICE output file is shown in Table 3.13.

When the analysis is performed, the first run is done using the nominal values of the devices. The results give the values for RB, RC, RE, and transistor beta for several runs. For each run, the value for VCC, node voltage V(2, 4), and current I(RC) are given. In addition, the summary of the Monte Carlo analysis shows the maximum value and its percentage change of the nominal value are given for the various runs.

3.8 Sensitivity and Worst-Case Analysis (.WCASE)

Critical elements in a circuit can be determined using WCASE statement. During the WCASE analysis, only one element is varied per run. This allows PSPICE to calculate the sensitivity of an output variable for each element parameter in the circuit. Once the sensitivity of each element is known, one final run is done with all the component parameters, this will give the worst case output.

TABLE 3.13

Edited Results of Monte Carlo Analysis

MONTE CARLO NOMINAL

**** CURRENT MODEL PARAMETERS FOR DEVICES REFERENCING RMOD

	RB	RC	RE
R	1.0000E + 00	1.0000E + 00	1.0000E + 00

**** CURRENT MODEL PARAMETERS FOR DEVICES REFERENCING QMOD

	Q1
BF	1.0000E + 02

MONTE CARLO ANALYSIS

VCC	V(2,4)	I(RC)
1.000E + 01	3.393E + 00	8.262E-03

****

MONTE CARLO ANALYSIS **MONTE CARLO PASS 2**

**** CURRENT MODEL PARAMETERS FOR DEVICES REFERENCING RMOD

	RB	RC	RE
R	9.5672E-01	9.6940E-01	1.0304E + 00

**** CURRENT MODEL PARAMETERS FOR DEVICES REFERENCING QMOD

	Q1
BF	9.0106E + 01

**** DC TRANSFER CURVES   TEMPERATURE = 27.000 DEG C

VCC	V(2,4)	I(RC)
1.000E + 01	3.931E + 00	7.989E-03

****

MONTE CARLO ANALYSIS **MONTE CARLO PASS 100**

**** UPDATED MODEL PARAMETERS   TEMPERATURE = 27.000 DEG C

**** CURRENT MODEL PARAMETERS FOR DEVICES REFERENCING RMOD

	RB	RC	RE
R	1.0143E + 00	1.0146E + 00	1.0410E + 00

**** CURRENT MODEL PARAMETERS FOR DEVICES REFERENCING QMOD

	Q1
BF	9.0247E + 01

**** DC TRANSFER CURVES   TEMPERATURE = 27.000 DEG C

VCC	V(2,4)	I(RC)
1.000E + 01	3.724E + 00	7.874E-03

****

TABLE 3.13

Edited Results of Monte Carlo Analysis (Continued)

```
MONTE CARLO ANALYSIS


 **** SORTED DEVIATIONS OF I(RC)    TEMPERATURE = 27.000 DEG C


                        MONTE CARLO SUMMARY
********************************************************************
RUN                     MAXIMUM VALUE

Pass 18                 8.6788E-03 at VCC = 10
                        ( 105.04% of Nominal)

Pass 31                 8.6637E-03 at VCC = 10
                        ( 104.86% of Nominal)

NOMINAL                 8.2623E-03 at VCC = 10

Pass 99                 8.2504E-03 at VCC = 10
                        ( 99.855% of Nominal)

Pass 61                 7.8950E-03 at VCC = 10
                        ( 95.554% of Nominal)

Pass 100                7.8736E-03 at VCC = 10
                        ( 95.295% of Nominal)
```

The component parameters that are varied in the sensitivity and worst-case analysis are specified by the DEV and LOT tolerances of each .MODEL parameter. This was discussed in Section 3.7. The general format for the .WCASE statement is:

.WCASE ANALYSIS OUTPUT_VARIABLE FUNCTION [OPTIONS]

where

.WCASE causes the sensitivity and worst-case analysis to be performed;

ANALYSIS is one of the following analysis types DC, AC, or TRAN. All the analysis types specified in the circuit file are performed during the first nominal run. Only the selected analysis, specified in the .WCASE statements are performed during the subsequent runs;

OUTPUT_VARIABLE is the output variable that is to be monitored. Its format is identical to that of the .PRINT output variable, discussed in Section 2.5.

FUNCTION specifies the operation to be performed on the output variable in order to reduce the values of the output variables to a single value. The function must be one of the following:

YMAX - finds the absolute value of the *greatest difference* in each waveform from the nominal run.

MAX - finds the maximum value of each waveform.
MIN - finds the minimum value of each waveform.

RISE_EDGE (< VALUE >) finds the *first occurrence* of the waveform crossing *above* the threshold < value >. The waveform must have one or
more points at or below < value > followed by one above. The output
value listed will be where the waveform increases above < value >.
FALL_EDGE < value > - finds the first occurrence of the waveform crossing
below the threshold < value >. The waveform must have one or more
points at or above < value > followed by one below. The output value
listed will be where the waveform decreases below < value >; and
OPTION includes none or more of the following:

LIST will print the updated model parameters for sensitivity
analysis.
OUTPUT ALL requests output from the sensitivity runs, after the nomi-
nal (first) run. The output from any run is governed by the .PRINT,
.PLOT, and .PROBE statements in the file. If OUTPUT ALL is omitted,
then only the nominal and worst-case runs produce output. OUTPUT
ALL will ensure that all sensitivity information is saved for PROBE.
RANGE (< low value >, < high value >) restricts the range over which
FUNCTION will be evaluated. An "X" can be used in place
of < value > to indicate for all values. If RANGE is omitted, then
FUNCTION is evaluated over the whole sweep range. This is equivalent to RANGE(*,*).
HI or LOW specifies what direction the worst-case run is to go (relative
to the nominal). If FUNCTION is YMAX or MAX, the default is HI.
Otherwise, the default is LOW.
VARY DEV/VARY LOT/VARY BOTH - by default, any device that has a
model parameter specifying either a DEV tolerance or a LOT tolerance will be included in the analysis. You may limit the analysis to
only those devices that have DEV or LOT tolerance by specifying the
appropriate option. The default is VARY BOTH. When VARY BOTH
is used, sensitivity to parameters with both DEV and LOT specifi-
cations is checked only with respect to LOT variations. The parameter is then maximized utilizing both DEV and LOT tolerances. All
devices referencing the model will have the same parameter value
for the worst-case simulation.
DEVICES (list of device types) by default, all devices are included in the
sensitivity and worst-case analysis. One can limit the devices considered by listing the device types after the keyword DEVICES.

It should be noted that < function > and all [options] do not affect PROBE
data obtained from the simulation. They are applicable to the output file.
The following example will explore the worst-case analysis statement for
two amplifier circuits.

Example 3.11: Worst-Case and Sensitivity Analysis of an Instrumentation Amplifier

For the instrumentation amplifier shown in Figure 3.20, R1 = 1 KΩ, R2 = R3 = 10 KΩ, R4 = R5 = 20 KΩ, and R6 = R7 = 100 KΩ. If the resistors have tolerance of 5%, find the sensitivity and the worst-case gain of the amplifier. The source voltage VIN has an amplitude of 1 mV and a frequency of 5 KHz. Assume the op amp has input resistance of 10^{12} Ohms, open loop gain of 10^7 and zero output resistance.

Solution

PSPICE Program

```
INSTRUMENTATION AMPLIFIER
.OPTIONS  RELTOL = 0.05; 5% COMPONENTS (SENSITIVITY RUN)
*
VIN 1   4   AC  1E-3;   INPUT SIGNAL
.AC LIN 10  1   5KHZ;   FREQUENCY OF SOURCE AND AC ANALYSIS
X1   2   1   5   OPAMP; OP AMP X1
X2   3   4   6   OPAMP; OP AMP X2
X3   8   7   9   OPAMP; OP AMP X3
*RESISTORS WITH MODELS
R1   2   3   RMOD1   1K
R2   2   5   RMOD1   10K
R3   3   6   RMOD1   10K
R4   5   8   RMOD1   20K
R5   6   7   RMOD1   20K
R6   7   0   RMOD1   100K
R7   8   9   RMOD1   100K
.MODEL RMOD1   RES(R = 1 DEV = 5%);  5% RESISTORS
*
.WCASE AC  V(9)   MAX  OUTPUT ALL;  SENSITIVITY & WORST CASE
*
.PROBE V(9)
*SUBCIRCUIT
.SUBCKT    OPAMP   1   2   3
* - INPUT; + INPUT; OUTPUT
RIN   1   2   1.0E12
EVO   0   3   1   2   1.0E7
.ENDS   OPAMP
.END
```

Table 3.14 shows the output file for the sensitivity analysis and Table 3.15 shows the worst-case results from the simulation.

In Table 3.14, the nominal value of the voltage V(9) at a frequency of 5 KHz is 0.105 V. Results for the output voltage are given for changes in R1 to R7. The output voltage is sorted with the maximum value of V(9) of 0.1098 V listed first (caused by the change in R7) and the minimum value of V(9) of 0.1002 V listed last (caused by the change in R1).

Table 3.15 shows that R1, R4, and R5 are set to their minimum allowed values and R2, R3, R6, and R7 are set their maximum allowed values. These values

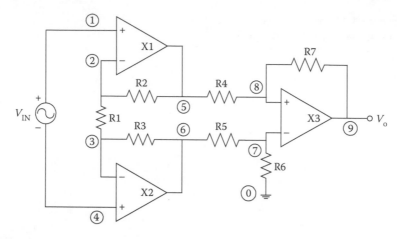

FIGURE 3.20
Instrumentation amplifier.

TABLE 3.14

Sensitivity Analysis of an Instrumentation Amplifier

INSTRUMENTATION AMPLIFIER

```
  **** SORTED DEVIATIONS OF V(9)    TEMPERATURE = 27.000 DEG C

                       SENSITIVITY SUMMARY

*************************************************************

   RUN          MAXIMUM VALUE

R7 RMOD1      R .1098 at F = 5.0000E + 03
              (.9167% change per 1% change in Model Parameter)

R2 RMOD1 R    .1075 at F = 5.0000E + 03
              (.4762% change per 1% change in Model Parameter)

R3 RMOD1 R    .1075 at F = 5.0000E + 03
              (.4762% change per 1% change in Model Parameter)

R6 RMOD1 R    .1054 at F = 5.0000E + 03
              (.08 % change per 1% change in Model Parameter)

NOMINAL       .105 at F = 5.0000E + 03

R5 RMOD1 R    .1046 at F = 5.0000E + 03
              (-.0826% change per 1% change in Model Parameter)

R4 RMOD1 R    .1004 at F = 5.0000E + 03
              (-.873 % change per 1% change in Model Parameter)

R1 RMOD1 R    .1002 at F = 5.0000E + 03
              (-.907 % change per 1% change in Model Parameter)
```

TABLE 3.15

Worst-Case Results

INSTRUMENTATION AMPLIFIER

```
 **** WORST CASE ANALYSIS          TEMPERATURE = 27.000 DEG C

                    WORST CASE ALL DEVICES

 *****************************************************************

  **** UPDATED MODEL PARAMETERS       TEMPERATURE = 27.000 DEG C

                    WORST CASE ALL DEVICES

 *****************************************************************

 DEVICE     MODEL     PARAMETER     NEW VALUE
 R1         RMOD1     R                .95 (Decreased)
 R2         RMOD1     R               1.05 (Increased)
 R3         RMOD1     R               1.05 (Increased)
 R4         RMOD1     R                .95 (Decreased)
 R5         RMOD1     R                .95 (Decreased)
 R6         RMOD1     R               1.05 (Increased)
 R7         RMOD1     R               1.05 (Increased)

 *****************************************************************

 **** SORTED DEVIATIONS OF V(9)      TEMPERATURE = 27.000 DEG C

                    WORST CASE SUMMARY

 *****************************************************************

  RUN                    MAXIMUM VALUE

 ALL DEVICES             .1277 at F = 5.0000E + 03
                         ( 121.61% of Nominal)

 NOMINAL                 .105 at F = 5.0000E + 03
```

condition the output voltage V(9) to be maximum. The maximum voltage at node 9, considering all devices, is 0.1277 V. This maximum value is 121.61% of the nominal value of 0.105 V.

Example 3.12: Worst-Case and Sensitivity Analysis of a Current-Biased Common Emitter Amplifier

For the common-emitter amplifier shown in Figure 3.21, R1 = 50 Ω, R2 = 1 KΩ, R3 = 10 KΩ, R4 = 1 KΩ, R5 = 16 KΩ, R6 = 20 KΩ, R7 = RL = 10 KΩ, VCC = 5 V, VEE = −5 V, CC1 = CC2 = 20 μF, and CE = 100 μF. Q1 and Q2 are transistors Q2N2222. The tolerance of the resistors is 5% and capacitors 5%. The transistor has gain BF of 100 with device variation of 30% (uniform distribution). What is the worst-case output voltage?

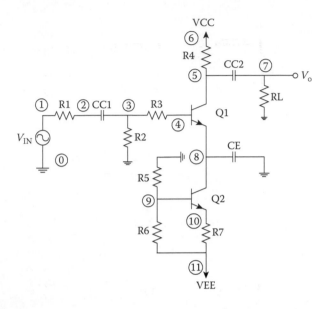

FIGURE 3.21
Current-biased common-emitter amplifier.

Solution

PSPICE Program

```
COMMON-EMITTER AMPLIFIER
.OPTIONS    RELTOL = 0.05;
VIN  1   0   AC  1E-3;  AC INPUT SIGNAL
*RESISTORS WITH MODEL
R1   1   2   RMOD2   50
R2   3   0   RMOD2   1K
R3   3   4   RMOD2   10K
R4   6   5   RMOD2   1K
R5   9   0   RMOD2   16K
R6   9   11  RMOD2   20K
R7   10  11  RMOD2   10K
RL   7   0   RMOD2   10K
.MODEL RMOD2   RES(R = 1 DEV = 5%); 5% RESISTORS
*
*CAPACITORS WITH MODEL
CC1  3   2   CMOD2   20E-6
CE   8   0   CMOD2   100E-6
CC2  5   7   CMOD2   20E-6
.MODEL CMOD2  CAP(C = 1 DEV = 5%); 5% CAPACITORS
*SOURCE VOLTAGES
VCC  6   0   DC      10V
VEE  11  0   DC      -10V
* TRANSISTOR WITH MODELS
Q1   5   4   8       Q2N2222
Q2   8   9   10      Q2N2222
```

```
.MODEL  Q2N2222  NPN  (BF = 100 DEV/UNIFORM 30% IS = 3.295E-14
VA = 200)
.AC LIN  1 1KHZ  1KHZ;  FREQUENCY OF SOURCE & AC ANALYSIS
* SENSITIVITY AND WORSE CASE ANALYSIS FOR AC ANALYSIS
.WCASE  AC  V(7))  MAX OUTPUT ALL; FOR AC ANALYSIS
.END
```

Table 3.16 shows edited sensitivity summary report. In addition, Table 3.17 shows the worst-case summary of results.

In Table 3.16, the nominal value of the voltage V(7) at a frequency of 1 KHz is 5.7756 mV. Results for the output voltage are given for changes in R1 to RL, CC1, CC2, CE, and also the changes in BF of transistors Q1 and Q2. The output voltage is sorted with the maximum value of V(7) being 6.0339 mV listed first (caused by the change in R4) and the minimum value of V(7) being 5.5953 mV listed last (caused by the change in R3).

Table 3.17 shows that CC1, CE, CC2, R1, R3, R5, R7, and BF of Q2 are set to their minimum allowed values and R2, R4, R6, RL, and BF of Q1 are set their maximum allowed values. These values condition the output voltage V(7) to be maximum. The maximum voltage at node 7, considering all devices, is 7.6507 mV. This maximum value is 132.47% of the nominal value of 5.7756 mV.

3.9 Fourier Series (.FOUR)

The periodic signal $g(t)$ can be expressed as an infinite series of sine and cosine terms, i.e.:

$$v(t) = \frac{a_0}{2} + \sum_{n=1}^{\infty} a_n \cos(nw_0 t) + b_n \sin(nw_0 t) \tag{3.12}$$

where

$$w_0 - \frac{2\pi}{T_P} \tag{3.13}$$

T_P = period of $v(t)$.

The Fourier coefficient a_n and b_n are determined by the following equations:

$$a_n = \frac{2}{T_P} \int_{t_0}^{t_0+T_P} v(t)\cos(n\omega_0 t)dt \quad n = 0,1,2,3... \tag{3.14}$$

TABLE 3.16

Edited Sensitivity Summary Report for Example 3.12

****

COMMON-EMITTER AMPLIFIER

**** SORTED DEVIATIONS OF V(7)   TEMPERATURE = 27.000 DEG C

                         SENSITIVITY SUMMARY

****

RUN   MAXIMUM VALUE

R4 RMOD2 R   6.0339E-03 at F = 1.0000E + 03
               (.8945% change per 1% change in Model Parameter)

Q1 Q2N2222 BF 5.9564E-03 at F = 1.0000E + 03
               (.626 % change per 1% change in Model Parameter)

R6 RMOD2 R   5.8245E-03 at F = 1.0000E + 03
               (.1693% change per 1% change in Model Parameter)

RL RMOD2 R   5.8012E-03 at F = 1.0000E + 03
               (.0885% change per 1% change in Model Parameter)

R2 RMOD2 R   5.7868E-03 at F = 1.0000E + 03
               (.0387% change per 1% change in Model Parameter)

NOMINAL   5.7756E-03 at F = 1.0000E + 03

Q2 Q2N2222 BF 5.7755E-03 at F = 1.0000E + 03
               (-490.2E-06% change per 1% change in Model
               Parameter)

CE CMOD2 C   5.7737E-03 at F = 1.0000E + 03
               (-6.5081E-03% change per 1% change in Model
               Parameter)

CC1 CMOD2 C   5.7737E-03 at F = 1.0000E + 03
               (-6.5516E-03% change per 1% change in Model
               Parameter)

CC2 CMOD2 C   5.7737E-03 at F = 1.0000E + 03
               (-6.6177E-03% change per 1% change in Model
               Parameter)

R1 RMOD2 R   5.7592E-03 at F = 1.0000E + 03
               (-.057 % change per 1% change in Model Parameter)

R5 RMOD2 R   5.7233E-03 at F = 1.0000E + 03
               (-.1813% change per 1% change in Model Parameter)

R7 RMOD2 R   5.6742E-03 at F = 1.0000E + 03
               (-.3513% change per 1% change in Model Parameter)

R3 RMOD2 R   5.5953E-03 at F = 1.0000E + 03
               (-.6244% change per 1% change in Model Parameter)

TABLE 3.17

Edited Worst-Case Summary for Example 3.12

```
******************************************************************
COMMON-EMITTER AMPLIFIER

 **** UPDATED MODEL PARAMETERS TEMPERATURE = 27.000 DEG C

 WORST CASE ALL DEVICES

******************************************************************
DEVICE     MODEL     PARAMETER     NEW VALUE
CC1        CMOD2     C                .95  (Decreased)
CE         CMOD2     C                .95  (Decreased)
CC2        CMOD2     C                .95  (Decreased)
Q1         Q2N2222   BF              130   (Increased)
Q2         Q2N2222   BF               70   (Decreased)
R1         RMOD2     R                .95  (Decreased)
R2         RMOD2     R               1.05  (Increased)
R3         RMOD2     R                .95  (Decreased)
R4         RMOD2     R               1.05  (Increased)
R5         RMOD2     R                .95  (Decreased)
R6         RMOD2     R               1.05  (Increased)
R7         RMOD2     R                .95  (Decreased)
RL         RMOD2     R               1.05  (Increased)
******************************************************************
                   WORST CASE ALL DEVICES
COMMON-EMITTER AMPLIFIER

 **** SORTED DEVIATIONS OF V(7)  TEMPERATURE = 27.000 DEG C

 WORST CASE SUMMARY

******************************************************************
 RUN              MAXIMUM VALUE
ALL DEVICES       7.6507E-03 at F = 1.0000E + 03
                  ( 132.47% of Nominal)

NOMINAL           5.7756E-03 at F - 1.0000E + 03
```

$$b_n = \frac{2}{T_P} \int_{t_0}^{t_0+T_p} v(t)\sin(n\omega_0 t)dt \quad n = 0,1,2,3... \tag{3.15}$$

Equation 3.12 can be rewritten as:

$$v(t) = C_0 + \sum_{n=1}^{\infty} c_n \sin(n\omega_0 t + \theta_n)dt \tag{3.16}$$

where

$$C_0 = \frac{a_0}{2} \qquad (3.17)$$

$$C_n = \sqrt{a_n^2 + b_n^2} \qquad (3.18)$$

and

$$\theta_n = \tan\left(\frac{a_n}{b_n}\right). \qquad (3.19)$$

C_0 is the DC component and C_n is the coefficient for the nth harmonic. The first harmonic is obtained with $n = 1$. The latter is also called the fundamental with fundamental frequency of w_0. When $n = 2$, we have the second harmonic, $n = 3$ the third harmonic, and so on. Fourier series representation in PSPICE is shown in Equation 3.16.

PSPICE allows the user to do Fourier Series analysis using the .FOUR statement. PSPICE calculates the signal Fourier components from DC to the nth component. The amplitude and phase of the harmonic components are tabulated in the circuit output file. The .FOUR statement results are printed automatically and it does not require .PRINT, .PLOT, or .PROBE statements. The general form of the .FOUR statements is:

.FOUR FUNDA_FREQUENCY NUMBER_OF_HARMONICS
OUTPUT_VARIABLE

where
  FUNDA_FREQUENCY is the fundamental frequency of the periodic waveform in Hertz;
  NUMBER_OF_HARMONICS is the number of harmonics to be calculated. The DC component, fundamental, and 2nd to 9th harmonics are calculated by default. More harmonics can be obtained by specifying the number of harmonics; and
  OUTPUT_VARIABLE is an output variable of the same form as in a .PRINT statement or .PLOT statement for transient analysis.

The .FOUR statement requires .TRAN statements. The Fourier series coefficients are obtained by PSPICE by performing a Fourier integral on selected output of transient analysis results at evenly spaced time points. The interval used is *print step size* in the .TRAN statement, or 1% of the *final time value*

(TSTOP) if smaller. A 2nd order polynomial interpolation is employed to calculate the transient analysis output values used in the integration. Not all the transient analysis results are used by the .FOUR statement. Only the interval from the end, back to 1/(fundamental frequency) before the end is used. This means that the transient-analysis must be at least 1/(fundamental frequency) seconds long. The following example illustrates the use of the .FOUR statement.

Example 3.13: Fourier Series Expansion of a Half-Wave Rectifier

Figure 3.22 shows a circuit for half-wave rectification. For a sinusoidal input with frequency of 500 Hz, amplitude of 1 V, find the frequency content at the output voltage, VO.

Solution

PSPICE Netlist

```
*HALF WAVE RECTIFIER
VS  1    0    SIN(0   1    500   0    0)
D1  1    2    DMOD
.MODEL   DMOD    D
RL  2    0    1K
*CONTROL STATEMENTS
.TRAN    1E-6    4E-3
.FOUR    500 V(2)
.END
```

The output of the program is shown in Table 3.18.
From the results of Table 3.16 , the output voltage can be written as

$$vo(t) = 0.073 + 0.133\sin(1000\pi t - 0.011°) + 0.098\sin(2000\pi t - 90.02°) + \quad (3.20)$$
$$0.055\sin(3000\pi t + 180°) + 0.019\sin(4000\pi t + 90.02°) + \ ...$$

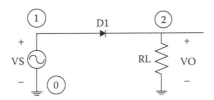

FIGURE 3.22
Half-wave rectifier.

TABLE 3.18

Fourier Components of an Output of a Half-Wave Rectifier

*HALF WAVE RECTIFIER

```
**** FOURIER ANALYSIS              TEMPERATURE = 27.000 DEG C
****************************************************************
FOURIER COMPONENTS OF TRANSIENT RESPONSE V(2)

DC COMPONENT = 7.341440E-02
```

HARMONIC NO	FREQUENCY (HZ)	FOURIER COMPONENT	NORMALIZED COMPONENT	PHASE (DEG)	NORMALIZED PHASE (DEG)
1	5.000E+02	1.333E-01	1.000E+00	-1.094E-02	0.000E+00
2	1.000E+03	9.815E-02	7.365E-01	-9.002E+01	-9.001E+01
3	1.500E+03	5.517E-02	4.140E-01	1.800E+02	1.800E+02
4	2.000E+03	1.881E-02	1.412E-01	9.002E+01	9.003E+01
5	2.500E+03	2.417E-03	1.814E-02	1.787E+02	1.787E+02
6	3.000E+03	8.469E-03	6.355E-02	8.953E+01	8.954E+01
7	3.500E+03	5.333E-03	4.002E-02	-2.022E-01	-1.913E-01
8	4.000E+03	2.029E-04	1.523E-03	-7.174E+01	-7.173E+01
9	4.500E+03	2.472E-03	1.855E-02	-2.385E+00	-2.374E+00

TOTAL HARMONIC DISTORTION = 8.603263E + 01 PERCENT

3.9.1 Fourier Analysis using PROBE

In the previous section, the Fourier series expansion was obtained in tabular form by using both .TRAN and .FOUR statements in the input file. The .OUT file contains the tabulated Fourier components. One can also use PROBE to obtain the Fourier components of a signal in a graphic format. This is achieved by:

- Having .TRAN and .PROBE statements as part of input file.
- Plotting using PROBE the transient analysis results of the required waveform.
- Selecting "X-axis" from PROBE menu
- Choosing "Fourier" from sub-menu

To reduce excessive data points saved to memory, it is important to specify the required output variables in the .PROBE statement. PROBE uses Fast Fourier transform (FFT) algorithm to obtain the Fourier components. FFT requires that: (1) the time intervals between the individual samples should be equally spaced, and (2) the number of data points should be a power of two. Before the Fourier components are calculated, PROBE creates a new set of data points (power of 2) based on the points created by PSPICE using interpolation method.

To illustrate the Fourier analysis using PROBE, we redo Example 3.13. The netlist is shown below.

```
*HALF WAVE RECTIFIER
VS   1    0    SIN(0    1    500    0    0)
D1   1    2    DMOD
.MODEL   DMOD    D
RL   2    0    1K
*CONTROL STATEMENTS
.TRAN    1E-6    4E-3
.FOUR    500  V(2)
.PROBE  V(2)
.END
```

The Fourier coefficients at output of the rectifier is displayed using PROBE. Select "x-axis" from the PROBE menu and then select "Fourier" from the sub-menu. The output obtained is shown in Figure 3.23.

3.9.2 RMS and Harmonic Distortion

Since the Fourier series expansion decomposes a periodic signal into the sum of sinusoids, the root-mean-squared (rms) value of the periodic signal can be obtained by adding the rms value of each harmonic component vectorially; that is,

$$V_{rms} = \sqrt{V_{1,\,rms}^2 + V_{2,\,rms}^2 + V_{3,\,rms}^2 + \;\cdots\; + V_{n,\,rms}^2} \tag{3.21}$$

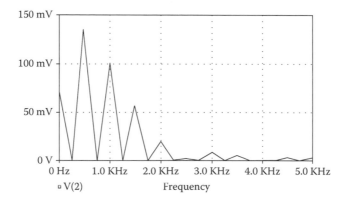

FIGURE 3.23
Frequency plot of half-wave rectifier.

where

V_{rms} = rms value of the periodic signal; and

$V_{1,rms}, V_{2,rms}, \ldots V_{n,rms}$ are rms values of the harmonic components.

From Equations 3.16 and 3.21, the rms value of $v(t)$ is:

$$V_{rms} = \sqrt{C_0^2 + \left(\frac{C_1}{\sqrt{2}}\right)^2 + \left(\frac{C_2}{\sqrt{2}}\right)^2 + \ldots \left(\frac{C_n}{\sqrt{2}}\right)^2}. \tag{3.22}$$

Harmonic distortion is a measure of the distortion a signal undergoes as it passes through a network (such as an amplifier, filter, transmission line). Harmonic distortion can also show the discrepancy between an approximation of a signal (obtained from synthesizing sinusoidal components) and its actual waveform. The smaller the harmonic distortion, the more nearly the approximation of a signal resembles the true signal. For the Fourier series expansion, the percent distortion for each individual component is given as:

$$\text{Percent distortion for } n\text{th harmonic} = \frac{C_n}{C_1} * 100 \tag{3.23}$$

where

C_n is the amplitude of the nth harmonic; and

C_1 is the amplitude of the fundamental harmonic.

The total harmonic distortion (THD) involves all the frequency components and it is given as:

$$\text{Percent THD} = \sqrt{\left(\frac{C_2}{C_1}\right)^2 + \left(\frac{C_3}{C_1}\right)^2 + \cdots \left(\frac{C_n}{C_1}\right)^2} * 100\% \tag{3.24}$$

or

$$\text{Percent THD} = \frac{\sqrt{C_2^2 + C_3^2 + \cdots C_n^2}}{C_1} * 100\%. \tag{3.25}$$

The following example determines the THD at the output of an RC network.

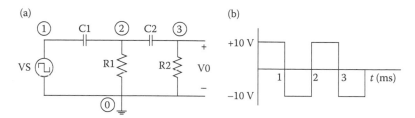

FIGURE 3.24
(a) A two-stage RC network, (b) input signal to RC network.

Example 3.14: Square Wave Signal Through Two-Stage RC Network

Figure 3.24a shows a two-stage RC network. C1 = C2 = 1 UF and R1 = R2 = 1 K. The periodic square wave, shown in Figure 3.24b, is applied to the input of RC network. Find the frequency content at the output. Determine the rms value of the output signal and calculate the THD of the output signal.

Solution

PSPICE Program

```
*SQUARE WAVE SIGNAL THROUGH NETWORK
VS      1      0
PULSE(-10   10   0   1E-6   1E-6   1E-3   2E-3)
C1      1      2           1E-6
R1      2      0           1E3
C2      2      3           1E-6
R2      3      0           1E3
.TRAN   5E-6     4E-3
.FOUR   500 V(1)  V(3)
.PROBE  V(1)     V(3)
.END
```

The Fourier coefficients for the first nine harmonics for output waveform voltage are shown in Table 3.19.

The rms value of the output voltage is obtained using Equation 3.22 and it is given as:

$$V_{rms} = \sqrt{(0.461)^2 + \left(\frac{9.608}{\sqrt{2}}\right)^2 + \left(\frac{0.0622}{\sqrt{2}}\right)^2 + \left(\frac{4.046}{\sqrt{2}}\right)^2 + \ldots \left(\frac{1.39}{\sqrt{2}}\right)^2}$$

$V_{rms} = 7.7588$ V.

The percent THD is given in Equation 3.24 and its value is:

$$\text{Percent THD} = \frac{\sqrt{(0.0622)^2 + (4.046)^2 + (0.0314)^2 + (2.484)^2 + \ldots (1.390)^2}}{9.608}$$

Percent THD = 54.74%.

The latter value agrees with that obtained from PSPICE simulation shown in Table 3.19.

The output waveform and its spectrum are shown in Figure 3.25.

TABLE 3.19

Fourier Components for Output Voltage of Figure 3.24

```
*SQUARE WAVE SIGNAL THROUGH NETWORK

 **** FOURIER ANALYSIS TEMPERATURE = 27.000 DEG C
 ***************************************************************
FOURIER COMPONENTS OF TRANSIENT RESPONSE V(3)

 DC COMPONENT = -4.608223E-01
```

HARMONIC NO	FREQUENCY (HZ)	FOURIER COMPONENT	NORMALIZED COMPONENT	PHASE (DEG)	NORMALIZED PHASE (DEG)
1	5.000E+02	9.608E+00	1.000E+00	4.678E+01	0.000E+00
2	1.000E+03	6.216E-02	6.470E-03	-1.736E+02	-2.204E+02
3	1.500E+03	4.046E+00	4.211E-01	1.655E+01	-3.023E+01
4	2.000E+03	3.145E-02	3.273E-03	-1.758E+02	-2.226E+02
5	2.500E+03	2.484E+00	2.585E-01	8.511E+00	-3.827E+01
6	3.000E+03	2.099E-02	2.184E-03	-1.761E+02	-2.229E+02
7	3.500E+03	1.784E+00	1.857E-01	4.470E+00	-4.231E+01
8	4.000E+03	1.573E-02	1.638E-03	-1.759E+02	-2.227E+02
9	4.500E+03	1.390E+00	1.447E-01	1.838E+00	-4.495E+01

```
TOTAL HARMONIC DISTORTION=5.473805E+01 PERCENT
```

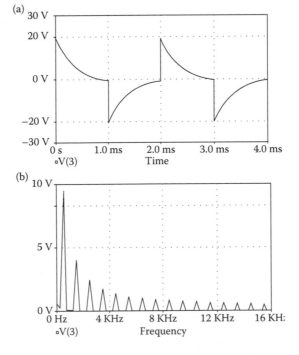

FIGURE 3.25
(a) Output waveform, (b) output spectra.

Problems

3.1   For the circuit shown in Figure P3.1, the resistance $R = 10 \text{ K}\Omega$.
(a) Find the change in the center frequency as C varies from
20 pF to 40 pF. (b) Plot the frequency response for the above
values of C.

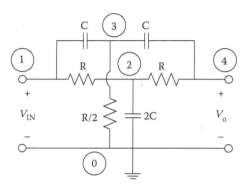

FIGURE P3.1
Twin-T network.

3.2   For the notched filter circuit shown in Figure P3.1, assume that
the capacitors and resistors are sensitive to temperature, find
the change in the notch frequency if the TC1 of the resistors and
capacitors is 1E-5. TC2 of resistors and capacitors is assumed to
be zero. $R = 10 \text{ K}\Omega$ and $C = 20$ pF.

3.3   For the voltage regulator circuit shown in Figure P3.3, VS =
20 V, RS = 300 Ω, RL = 4 KΩ, and D1 is D1N4742. Find the output
voltage as a function of temperature as the latter varies from
25°C to 50°C.

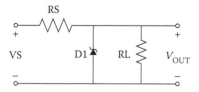

FIGURE P3.3
Voltage regulator.

3.4   For the voltage multiplier circuit shown in Figure 3.12a, the
waveform shown in Figure P3.4 is connected to the two input
of the multiplier, find the output voltage.

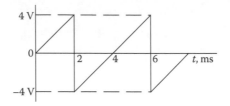

FIGURE P3.4
Input Waveform.

3.5   An amplitude modulated wave is given as $s(t) = A_c \cos(2\pi f_c t + k_a m(t))$.

If $m(t) = \cos(2\pi f_m t)$, $f_m = 10^4$ Hz, $f_c = 10^6$ Hz; $k_a = 0.5$; and $A_c = 10$ V, sketch $m(t)$ and $s(t)$ using the analog behavioral model technique.

3.6   A zener diode in the voltage regulator circuit of Figure P3.6 has corresponding current and voltage relationships shown in Table P3.6. R1 = R3 = R4 = 5 KΩ and R2 = 25 KΩ. Find the output voltage of the voltage regulator circuit when input changes from 20 V to 25 V.

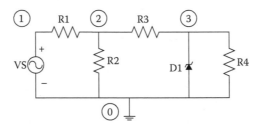

FIGURE P3.6
Voltage regulator circuit.

TABLE P3.6

Zener Diode Characteristics

Reverse Voltages, V	Reverse Current, A
1	1.0e–11
3	1.0e–11
4	1.0e–10
5	1.0e–9
6	1.0e–7
7	1.0e–6
7.5	2e–6
7.7	15.0e–4
7.9	44.5e–4

3.7 Figure P3.7 shows a cascaded system with subsystems A, B, and C. The frequency response data of each of the cascaded systems is shown in Table P3.7. Plot the frequency magnitude responses at the output of circuit A and C. What is bandwidth of the complete system?

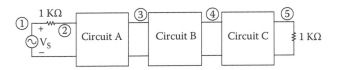

FIGURE P3.7
Cascaded circuit.

TABLE P3.7

Frequency Response of Each of the Subsystem A, B, and C

Frequency, Hz	Gain of Circuit A (dB)	Gain of Circuit B (dB)	Gain of Circuit C (dB)
1000	10.0	0.8	2
2000	9.5	1.2	2
3000	8.3	1.5	2
4000	6.5	2.0	2
5000	5.0	2.5	2
6000	4.0	3.0	2
7000	3.0	4.0	2
8000	2.5	5.0	2
9000	2.0	6.5	2
10000	1.5	8.3	2
11000	1.2	9.5	2
12000	0.8	10.0	2

3.8 In Example 3.14, calculate the rms value of the input periodic waveform. What is the percent THD of the input waveform?

3.9 A system has Laplace transform given as:

$$\frac{V_{OUT}}{V_{IN}}(s) = \frac{s^2 + 2s + 3}{s^3 + 9s^2 + 15s + 25}.$$

Assuming the input is a step voltage with amplitude of 5 V, determine the response of the system.

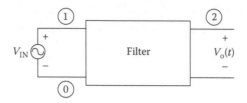

FIGURE P3.9
System with Laplace transform.

3.10 The open loop gain of a frequency compensated op amp is given as:

$$\frac{V_{OUT}}{V_{IN}}(s) = \frac{A_0}{1 + s/(2\pi f_p)}.$$

For the circuit shown in Figure P3.10, R1 = 1 KΩ, R2 = 2 KΩ, R3 = 4 KΩ, and R4 = 9 KΩ. If the op amp is frequency compensated with $A_0 = 10^5$, $f_p = 10$ Hz, plot the frequency response of V_{OUT}/V_{IN}.

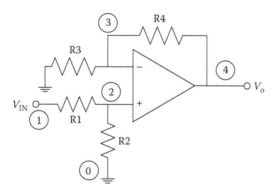

FIGURE P3.10
An operational amplifier circuit.

3.11 For the Sallen–Key filter circuit shown in Figure P3.11, R1 = R2 = R3 = 10 KΩ, R4 = 40 KΩ, C1 = C2 = 0.02 μF, VCC = 15 V, and VEE = −15 V. If resistors and capacitors have 5% tolerance values, calculate the range of output voltage from the nominal value for 25, 50, 100, and 125 runs. Assume that a 741 op amp is used, and the source has a frequency of 50 HZ and an amplitude of $1V_o$.

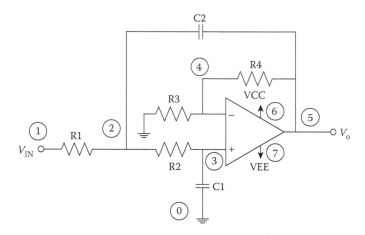

FIGURE P3.11
Sallen–Key filter circuit.

3.12 For the 5th order low pass filter, R1 = R2 = R3 = R4 = R5 = 1 KΩ and C1 = C2 = C3 = C4 = C5 = 1 μF. (a) If the resistors and capacitors have tolerances of 5%, calculate the range of values of VO from its nominal value. Use 50 runs. (b) If the resistors and capacitors have tolerances of 10%, calculate the range of values of VO from its nominal values. Use 50 runs.

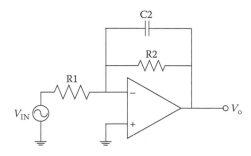

FIGURE P3.12
5th order low pass filter.

3.13 The Miller Integrator with gain at DC is shown in Figure P3.13. If the tolerance on the resistors is 5% and capacitors 5%, find the worst case output voltage for frequencies between 10 Hz

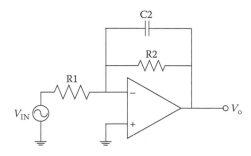

FIGURE P3.13
Miller integrator with DC gain.

to 20 KHz. Assume that the op amp has an input resistance of 10^{10} Ω, zero output resistance and open loop gain of 10^8.

3.14 For the amplifier shown in Figure P3.14, RS = 150 Ω, RB2 = 20 KΩ, RB1 = 90 KΩ, RE = 2 KΩ, RC = 5 KΩ, RL = 10 KΩ, C1 = 2 μF, CE = 50 μF, C2 = 2 μF, and VCC = 15 V. If the resistors have a tolerance of 5% and the capacitors 5%, determine the sensitivity and the worst case quiescent operating point of the transistor if the gain BF of transistor is 100 with variation of 40% (uniform distribution). Assume that Q1 is Q2N2222 transistor.

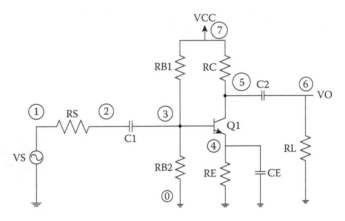

FIGURE P3.14
An amplifier circuit.

3.15 A full wave rectifier is shown in Figure P3.15. RL = 10 K, D1, D2, D3, and D4 are D1N4009 diodes. For this example, the input sinusoidal signal has a peak amplitude of 10 V and frequency is 60 Hz. (a) Find the Fourier series coefficients of the output load. (b) What is the rms value of the load voltage? (c) Calculate the power delivered to the load. (d) If a 10 μF capacitor is connected across the load resistor, use PROBE to display the frequency components of the output voltage.

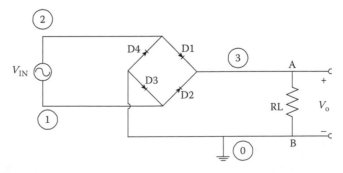

FIGURE P3.15
Full-wave rectifier circuit.

3.16 A saw-tooth waveform, shown in Figure P3.16b is applied at an input of an RC network, shown in Figure P3.16a. R1 = 2 KΩ and c1 = 5 μF. (a) Find the frequency components of a saw-tooth waveform. (b) Determine the THD of the voltage across the capacitor. (c) What is the rms value of the voltage across the capacitor?

(a)   (b)

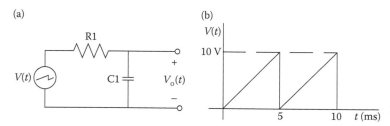

FIGURE P3.16
(a) RC network. (b) Input saw-tooth waveform.

3.17 A sinusoidal signal $V_s(t)$ is applied at the input of an RLC circuit. R1 = 2 KΩ, L1 = 3 mH, and C1 = 0.1 μF. If $V_s(t) = 10\sin(2000\pi t)$ volts, determine the frequency content of $V_o(t)$. What is the THD at the output?

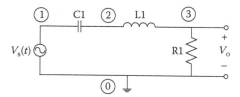

FIGURE P3.17
RLC circuit.

3.18 For the 4th high pass filter, R1 = R2 = R3 = R4 = 2 KΩ and C1 = C2 = C3 = C4 = 0.01 μF. (a) If the resistors and capacitors

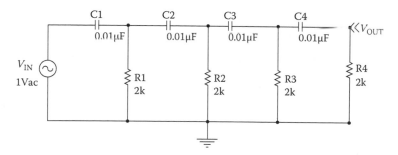

FIGURE P3.18
4th order low pass filter.

have tolerances of 5%, calculate the range of values of V_{OUT} from its nominal value. Use 50 runs. (b) If the resistors and capacitors have tolerances of 10%, calculate the range of values of V_{OUT} from its nominal values. Use 50 runs.

3.19   A system has Laplace transform given as:

$$\frac{V_{OUT}}{V_{IN}}(s) = \frac{16s}{s^2 + 9s + 20}.$$

Assuming the input is a step voltage with amplitude of 2 V, determine the response of the system.

3.20   A system has Laplace transform given as:

$$\frac{V_{OUT}}{V_{IN}}(s) = \frac{10^4}{s^2 + 200s + 10^6}.$$

Assuming the input is a transient sinusoid given as $v_{IN}(t) = 10\sin(10^6 \pi t)$, determine $v_{OUT}(t)$.

Bibliography

1. Al-Hashimi, Bashir. *The Art of Simulation Using PSPICE, Analog, and Digital.* Boca Raton, FL: CRC Press, 1994.
2. Alexander, Charles K., and Matthew N. O. Sadiku. *Fundamentals of Electric Circuits.* 4th ed. New York: McGraw-Hill, 2009.
3. Attia, J. O. *Electronics and Circuit Analysis Using MATLAB®.* 2nd ed. Boca Raton, FL: CRC Press, 2004.
4. Connelly, J. Alvin, and Pyung Choi. *Macromodeling with SPICE.* Upper Saddle River, NJ: Prentice Hall, 1992.
5. Conrad, William R. "Solving Laplace Transform Equation Using PSPICE." *Computers in Education Journal,* Vol. 5, no. 1 (January–March 1995): 35–37.
6. Distler, R. J. "Monte Carlo Analysis of System Tolerance." *IEEE Transactions on Education,* Vol. 20 (May 1997): 98–101.
7. Ellis, George. "Use SPICE to Analyze Component Variations in Circuit Design," *Electronic Design News (EDN)* (April 1993): 109–14.
8. Eslami, Mansour, and Richard S. Marleau. "Theory of Sensitivity of Network: A Tutorial." *IEEE Transactions on Education,* Vol. 32, no. 3 (August 1989): 319–34.
9. Fenical, L. H. *PSPICE: A Tutorial.* Upper Saddle River, NJ: Prentice Hall, 1992.
10. Hamann, J. C, Pierre, J. W., Legowski, S. F. and Long, F. M. "Using Monte Carlo Simulations to Introduce Tolerance Design to Undergraduates," IEEE Trans. On Education, Vol. 42, no. 1, pp. 1–14, (February 1999).
11. Hart, Daniel W. "Introducing Fourier Series Using PSPICE Computer Simulation." *Computers in Education, Division of ASEE* III, no. 2 (April–June 1993): 46–51.

12. Howe, Roger T., and Charles G. Sodini. *Microelectronics, An Integrated Approach.* Upper Saddle River, NJ: Prentice Hall, 1997.
13. Kavanaugh, Micheal F. "Including the Effects of Component Tolerances in the Teaching of Courses in Introductory Circuit Design." *IEEE Transactions on Education,* Vol. 38, no. 4 (November 1995): 361–64.
14. Keown, John. *PSPICE and Circuit Analysis.* New York: Maxwell Macmillan International Publishing Group, 1991.
15. Kielkowski, Ron M. *Inside SPICE, Overcoming the Obstacles of Circuit Simulation.* New York: McGraw-Hill, Inc., 1994.
16. Lamey, Robert. *The Illustrated Guide to PSPICE.* Clifton Park, NY: Delmar Publishers Inc., 1995.
17. Nilsson, James W., and Susan A. Riedel. *Introduction to PSPICE Manuel Using ORCAD Release 9.2 to Accompany Electric Circuits.* Upper Saddle River, NJ: Pearson/Prentice Hall, 2005.
18. OrCAD Family Release 9.2. San Jose, CA: Cadence Design Systems, 1986–1999.
19. Rashid, Mohammad H. *Introduction to PSPICE Using OrCAD for Circuits and Electronics.* Upper Saddle River, NJ: Pearson/Prentice Hall, 2004.
20. Roberts, Gordon W., and Adel S. Sedra. *Spice for Microelectronic Circuits.* Philadelphia, PA: Saunders College Publishing, 1992.
21. Sedra, A. S., and K. C. Smith. *Microelectronic Circuits.* 4th ed. Oxford: Oxford University Press, 1998.
22. Spence, Robert, and Randeep S. Soin. *Tolerance Design of Electronic Circuits.* London: Imperial College Press, 1997.
23. Soda, Kenneth J. "Flattening the Learning Curve for ORCAD-CADENCE PSPICE," *Computers in Education Journal,* Vol. XIV (April–June 2004): 24–36.
24. Svoboda, James A. *PSPICE for Linear Circuits.* 2nd ed. New York: John Wiley & Sons, Inc., 2007.
25. Thorpe, Thomas W. *Computerized Circuit Analysis with Spice.* New York: John Wiley & Sons, Inc., 1991.
26. Tobin, Paul. "The Role of PSPICE in the Engineering Teaching Environment." Proceedings of International Conference on Engineering Education, Coimbra, Portugal, September 3–7, 2007.
27. Tobin, Paul. *PSPICE for Circuit Theory and Electronic Devices.* San Jose, CA: Morgan & Claypool Publishers, 2007.
28. Tront, Joseph G. *PSPICE for Basic Circuit Analysis.* New York: McGraw-Hill, 2004.
29. Tuinenga, Paul W. *SPICE, A Guide to Circuit Simulations and Analysis Using PSPICE.* Upper Saddle River, NJ: Prentice Hall, 1988.
30. *Using MATLAB, The Language of Technical Computing, Computation, Visualization, Programming, Version 6.* Natick, MA: MathWorks, Inc., 2000
31. Vladimirescu, Andrei. *The Spice Book.* New York: John Wiley and Sons, Inc., 1994.
32. Wyatt, Michael A. "Model Ferrite Beads in SPICE." In *Electronic Design,* October 15, 1992.
33. Yang, Won Y., and Seung C. Lee. *Circuit Systems with MATLAB® and PSPICE®.* New York: John Wiley & Sons, 2007.

Part II

4

MATLAB® Fundamentals

MATLAB® is a numeric computation software for engineering and scientific calculations. The name MATLAB stands for MATRIX LABORATORY. MATLAB is primarily a tool for matrix computations. MATLAB has a rich set of plotting capabilities. The graphics are integrated in MATLAB. Since MATLAB is also a programming environment, a user can extend the functional capabilities to guarantee high accuracy. This chapter will introduce the basic operations of MATLAB, the control statements and plotting functions.

4.1 MATLAB® Basic Operations

When MATLAB® is invoked, the command window will display the prompt >>. MATLAB is then ready for entering data or executing commands. To quit MATLAB, type the command:

exit or quit

MATLAB has on-line help. To see the list of MATLAB's help facility, type:

Help

The help command followed by function names is used to obtain information on a specific MATLAB function. For example, to obtain information on the use of fast Fourier transform function, **fft,** one can type the command:

help fft

The basic data object in MATLAB is a rectangular numerical matrix with real or complex elements. Scalars are thought of a 1-by-1 matrix. Vectors are considered as matrices with a row or column. MATLAB has no dimension statement or type declarations. Storage of data and variables is allocated automatically once the data and variables are used.

MATLAB statements are normally of the form:

variable = expression

Expressions typed by the user are interpreted and immediately evaluated by the MATLAB system. If a MATLAB statement ends with a semicolon, MATLAB evaluates the statement but suppresses the display of the results. A matrix:

$$A = \begin{bmatrix} 6 & 7 & 8 \\ 9 & 10 & 11 \\ 12 & 13 & 14 \end{bmatrix}$$

may be entered as follows,

```
A = [6   7   8;   9   10  11;  12   13   14];
```

Note that the matrix entries must be surrounded by brackets [ ] with row elements separated by blanks or by commas. A semicolon indicates the end of each row, with the exception of the last row. A matrix A can also be entered across three input lines as:

```
A = [6   7   8
     9  10  11
    12  13  14];
```

In this case, the carriage returns replace the semicolons. A row vector B with four elements:

B = [30   40   60   90   71]

can be entered in MATLAB as

```
B = [30  40  60   90   71];
```

or

```
B = [30, 40, 60,  90,  71];
```

For readability, it is better to use spaces rather than commas between the elements. The row vector B can be turned into a column vector by transposition, which is obtained by typing:

C = B′

The above results in

```
C =
    30
    40
    60
    90
    71
```

TABLE 4.1

Some Basic MATLAB® Commands

Command	Description
%	Comments. Everything appearing after % command is not executed
demo	Access on-line demo programs
length	Length of matrix
clear	Clears the variables or functions from workspace
clc	Clears the command window during a work session
clg	Clears graphic window
diary	Saves a session in a disk, possibly for printing at a later date

Other ways of entering the column vector C are:

```
C = [30
     40
     60
     90
     71]
```

or

```
C = [30; 40; 60; 90; 71]
```

MATLAB is case sensitive in naming variables, commands, and functions. Thus **b** and **B** are not the same variables. If you do not want MATLAB to be case sensitive, you must use the command:

casesen off

Table 4.1 shows additional MATLAB commands to get one started on MATLAB. Detailed descriptions and usages of the commands can be obtained from the MATLAB help facility or from MATLAB manuals.

4.2 Matrix Operations

The basic matrix operations are addition (+), subtraction (−), multiplication (*), and conjugate transpose (') of matrices. In addition to the above basic operations, MATLAB® has two forms of matrix division: the left inverse operator \ or the right inverse operator /.

Matrices of the same dimension may be **subtracted or added**. Thus if E and F are entered in MATLAB as:

```
E = [21    25    30;    7    18    34;    70    16    8];
F = [1      7     3;    8    11     4;     2    11   13];
```

and

```
G = E - F
H = E + F
```

then, matrices G and H will appear on the screen as

```
G =
      20    18    27
      -1     7    30
      68     5    -5


H =
      22    32    33
      15    29    38
      72    27    21
```

Matrix multiplication is defined provided the inner dimensions of the two operands are the same. Thus, if X is an n-by-m matrix and Y is i-by-j matrix, X*Y is defined provided m is equal to i. Since matrices E and F are 3-by-3 matrices, the product:

```
Q = E*F
```

results as

```
Q =
      281    752    553
      219    621    535
      214    754    378
```

Any matrix can be multiplied by a scalar. For example,

```
2*Q
```

gives

```
ans =
         562    1504    1106
         438    1242    1070
         428    1508     756
```

TABLE 4.2

Some Common MATLAB Functions

Functions	Description
abs(x)	Calculates the absolute value of x
acos(x)	Determines $\cos^{-1}x$, with the results in radians
asin(x)	Determines $\sin^{-1}x$ with the results in radians
atan(x)	Calculates $\tan^{-1}x$, with the results in radians
atan2(x)	Obtains $\tan^{-1}(y/x)$ over all four quadrants of the circle. The results are in radians.
cos(x)	Calculates cos (x), with x in radians
exp(x)	Computes e^x
log(x)	Determines the natural logarithm $\log_e(x)$
sin(x)	Calculates sin(x), with x in radians
sqrt(x)	Computes the square root of x
tan(x)	Calculates tan(x), with x in radians

Note that if a variable name and the "=" sign are omitted; a variable name **ans** is automatically created.

Matrix division can either be the **left division operator** \ or the **right division operator** /. The right division **a/b**, for instance, is algebraically equivalent to a/b while the left division a\b is algebraically equivalent to b/a.

If $Z^*I = V$ and Z is nonsingular, the left division, $Z\backslash V$ is equivalent to MATLAB expression:

$$I = \text{inv}(Z)^*V$$

where

inv is the MATLAB function for obtaining the inverse of a matrix.

The right division denoted by V/Z is equivalent to the MATLAB expression:

$$I = V^* \, \text{inv}(Z) \tag{4.1}$$

Apart form the function **inv**, there are additional MATLAB functions that are worth noting. They are given in Table 4.2.

The following example uses the inv function to determine the nodal voltages of a resistive circuit.

Example 4.1: Nodal Analysis of a Resistive Network

For the circuit shown in Figure 4.1, the resistances are in Ohms. Write the nodal equations and solve for voltages V_1, V_2, and V_3.

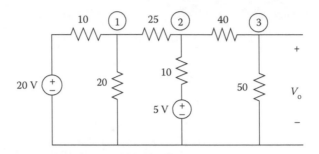

FIGURE 4.1
Resistive network.

Solution

Using KCL and assuming that currents leaving a node are positive, we have, for node 1:

$$\frac{V_1 - 20}{10} + \frac{V_1}{20} + \frac{V_1 - V_2}{25} = 0.$$

Simplifying, we have:

$$0.19V_1 - 0.04V_2 = 2. \tag{4.2}$$

For node 2:

$$\frac{V_2 - V_1}{25} + \frac{V_2 - 5}{10} + \frac{V_2 - V_3}{40} = 0.$$

Simplifying, we have:

$$-0.03V_1 + 0.165V_2 - 0.025V_3 = 0.5 \tag{4.3}$$

For node 3:

$$\frac{V_3}{50} + \frac{V_3 - V_2}{40} = 0.$$

Simplifying, we have:

$$-0.025V_2 + 0.045V_3 = 0. \tag{4.4}$$

In matrix form, Equations 4.2 to 4.4 become:

$$\begin{bmatrix} 0.19 & -0.04 & 0 \\ -0.04 & 0.165 & -0.025 \\ 0 & -0.025 & 0.045 \end{bmatrix} \begin{bmatrix} V_1 \\ V_2 \\ V_3 \end{bmatrix} = \begin{bmatrix} 2 \\ 0.5 \\ 0 \end{bmatrix}$$

The MATLAB® function **inv** is used to obtain the nodal voltages. The MATLAB for solving the nodal voltages is:

MATLAB Script

```
% This program computes the nodal voltages
% given the admittance matrix Y and current vector I
% Y is the admittance matrix
% I is the current vector
% Initialize the matrix y and vector I using YV = I
Y = [0.19    -0.04     0;
      -0.04    0.165   -0.025;
       0      -0.025    0.045];
% current is entered as a transpose of row vector
I = [2    0.5    0]';
fprintf('Nodal voltages V1, V2, and V3 are \n')
V = inv(Y)*I
```

We obtain the following results.
Nodal voltages V1, V2, and V3 are:

V =
   11.8852
    6.4549
    3.5861

4.3 Array Operations

Array operations refer to element-by-element arithmetic operations. Preceding the linear algebraic matrix operations, * / \ ', by a period (.) indicates an array or element-by-element operation. Thus, the operators, .*, .\, ./, .^, represent element-by-element multiplication, left division, right division, and raising to the power, respectively. For addition and subtraction, the array and matrix operations are the same.

If K1 and K1 matrices are of the same dimensions, then A1.*B1 denotes an array whose elements are products of the corresponding elements of A1 and B1. Thus, if:

```
K1 = [1    7    4
      2    5    6];

L1 = [11   12   14
       7    4    1];
```
then

```
M1 = K1.*L1
```

results in

```
M1 =
      11   84   56
      14   20    6
```

An array operation for left and right division also involves element-by-element operation. The expressions **K1./L1 and K1.\L1** give the quotient of element-by-element division of matrices K1 and L1. The statement:

```
N1 = K1./L1
```

gives the result

```
N1 =
      0.0909  0.5833  0.2857
      0.2857  1.2500  6.0000
```

and the statement

P1 = K1.\L1

gives

```
P1 =
      11.0000  1.7143  3.5000
       3.5000  0.8000  0.1667
```

Array exponentiation is denoted by .^. The general statement will be of the form:

q = r1.^s1.

If r1 and s1 are matrices of the same dimensions, then the result q1 is also a matrix of the same dimensions. For example, if:

```
r1 = [4   3   7];
s1 = [1   4   3];
```

then

q1 = r1.^s1

gives the result

```
q1 =
       4  81  343
```

One of the operands can be scalar. For example,

```
q2 = r1.^2
q3 = (2).^s1
```

will give

```
q2 =
      16   9   49
```

and

```
q3 =
       2   16   8
```

Note that when one of the operands is scalar, the resulting matrix will have the same dimensions as the matrix operand.

4.4 Complex Numbers

MATLAB® allows operations involving complex numbers. Complex numbers are entered using function **i or j**. For example, a number $Z = 5 + j12$ may be entered in MATLAB as:

$$z = 5 + 12* \, i$$

or

$$z = 5 + 12* \, j$$

Also, a complex number $z1$

$$z1 = 4\sqrt{3}\exp\left[\left(\frac{\pi}{3}\right)j\right]$$

can be entered in MATLAB as:

$$z1 = 4*sqrt(3)*\exp\left[\left(\frac{pi}{3}\right)*j\right]$$

It should be noted that when complex numbers are entered as matrix elements within brackets, one should avoid any blank spaces. For example, z2 = 5 + j12 is represented in MATLAB as:

$$z2 = 5 + 12*j$$

If spaces exist around the + sign, such as:

$$z3 = 5 + 12*j$$

MATLAB considers it as two separate numbers, and z2 will not be equal to z3.

If y is a complex matrix given as:

$$y = \begin{bmatrix} 1+j1 & 2-j2 \\ 3+j2 & 4+j3 \end{bmatrix}$$

then we can represent it in MATLAB as

$$y = [1+j \quad 2-2*j; 3+2*j \quad 4+3*j]$$

which will produce the result

```
y =
      1.0000 + 1.0000i    2.0000 - 2.0000i
      3.0000 + 2.0000i    4.0000 + 3.0000i
```

If the entries in a matrix are complex, then the "prime" (') operator produces the conjugate transpose. Thus,

yp = y'

will produce

```
yp =
      1.0000 - 1.0000i    3.0000 - 2.0000i
      2.0000 + 2.0000i    4.0000 - 3.0000i.
```

For the unconjugate transpose of a complex matrix, we can use the point transpose (.') command. For example,

yt = y.'

will yield

TABLE 4.3

Some MATLAB Functions for Manipulating Complex Numbers

Function	Description
conj(Z)	Obtains the complex conjugate of a number Z. If Z = x + iy, then Conj(Z) = x − iy.
real(Z)	Returns the real part of the complex number Z
imag(Z)	Returns the imaginary part of the complex number Z
abs(Z)	Computes the magnitude of the complex number Z
angle(Z)	Calculates the angle of the complex number Z, determined from the expression atan2(imag(Z), real(Z)).

```
yt =
      1.0000 + 1.0000i    3.0000 + 2.0000i
      2.0000 - 2.0000i    4.0000 + 3.0000i
```

There are several functions for manipulating complex numbers. Some of the functions are shown in Table 4.3.

Example 4.2: Input Impedance of Oscilloscope PROBE

A simplified equivalent circuit of an oscilloscope PROBE for measuring low frequency signals is shown in Figure 4.2. If R1 = 9 MΩ, R2 = 1 MΩ, C1 = 10 pF, and C2 = 100 pF. What is the input impedance at a sinusoidal frequency of 20 KHz?

Solution

The input impedance is:

$$Z_{IN} = \left[\frac{1}{j\omega C_1} \middle\| R_1 \right] + \left[\frac{1}{j\omega C_2} \middle\| R_2 \right] = \frac{R_1}{1 + j\omega C_1 R_1} + \frac{R_2}{1 + j\omega C_2 R_2} \tag{4.5}$$

MATLAB® is used to evaluate Z_{IN} for various values of frequency w.
   The MATLAB program:

```
% ZIN is input impedance
c1 = 10e-12; c2 = 100e-12;
r1 = 9e+6;          r2 = 1.0e+6;
w = 2*pi*20.0e + 3;
z1 = 1 + j*w*c1*r1;
z2 = 1 + j*w*c2*r2;
zin = (r1/z1) + (r2/z2);
zin
```

The solution obtained is:

```
zin =
      7.6109e + 004 -8.6868e + 005i.
```

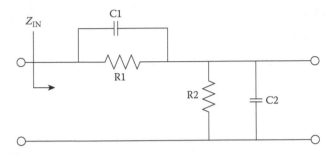

FIGURE 4.2
Simplified equivalent circuit of oscilloscope PROBE.

4.5 The Colon Symbol

The colon symbol (:) is one of the most important operators in MATLAB®. It can be used (a) to create vectors and matrices, (b) to specify sub-matrices and vectors, and (c) to perform iterations.

(a) *Creation of vectors and matrices*

The statement:

```
j1 = 1:8
```

will generate a row vector containing the numbers from 1 to 8 with unit increments. MATLAB produces the result:

```
j1 =
      1   2   3   4   5   6   7   8
```

Nonunity, positive, or negative increments may be specified. For example, the statement:

```
j2 = 4:-0.5:1
```

will yield the result

```
j2 =
      4.0000  3.5000  3.0000  2.5000  2.0000  1.5000 1.0000
```

The statement:

```
j3 = [(0:2:10); (5:-0.2:4)]
```

will result in a 2-by-4 matrix

```
j3 =
    0        2.0000   4.0000   6.0000   8.0000   10.0000
    5.0000   4.8000   4.6000   4.4000   4.2000    4.0000
```

(b) *Specifying submatrices and vectors*

Individual elements in a matrix can be referenced with subscripts inside parentheses. For example, j2(4) is the fourth element of vector j2. Also, for matrix j3, j3(2, 3) denotes the entry in the second row and third column. Using the colon as one of the subscripts denotes all of the corresponding row or column. For example, j3(:,4) is the fourth column element of matrix j3. Thus, the statement:

```
j5 = j3(:, 4)
```

will give

```
j5 =
    6.0000
    4.4000
```

Also, the statement j3(2, :) is the second row of matrix j3. That is the statement:

```
j6 = j3(2,:)
```

will result in

```
j6 =
    5.0000   4.8000   4.6000   4.4000   4.2000   4.0000
```

If the colon exists as the only subscript, such as j3(:), the latter denotes the elements of matrix j3 strung out in a long column vector. Thus, the statement:

```
j7 = j3(:)
```

will result in

```
j7 =
    0
    5.0000
    2.0000
    4.8000
    4.0000
    4.6000
    6.0000
    4.4000
```

```
8.0000
4.2000
10.0000
4.0000
```

(c) *Iterative uses of colon command*

The iterative uses of the colon command are discussed in the next section.

4.6 FOR Loops

"**FOR**" **loops** allow a statement or group of statements to be repeated a fixed number of times. The general form of a **for loop** is:

for index = expression
      statement group C
end

The expression is a matrix and the statement group C is repeated as many times as the number of elements in the columns of the expression matrix. The index takes on the elemental values in the matrix expression. Usually, the expression is something like:

m:n or m:i:n

where **m** is the beginning value, **n** the ending value, and **i** is the increment.

Suppose we would like to find the cubes of all the integers starting from 1 to 50. We could use the following statement to solve the problem:

```
sum = 0;
for i = 1:50
    sum = sum + i^3;
end
sum
```

FOR loops can be nested, and it is recommended that the loop be indented for readability. Suppose we want to fill a 5-by-6 matrix, *a*, with an element value equal to unity, the following statements can be used to perform the operation:

```
%
n = 5;              %number of rows
m = 6;              %number of columns
for i = 1:n
    for j = 1:m
        a(i,j) = 1;   %semicolon suppresses printing in the loop
    end
```

```
end
a                 %display the result
%
```

It is important to note that each for statement group must end with the word **end**. The following example illustrates the use of the **for** loop.

Example 4.3: Frequency Response of a Notched Filter

A notched filter eliminates a small band of frequencies. It has a transfer function given as:

$$H(s) = \frac{k_p(s^2 + \omega_0^2)}{s^2 + \left(\dfrac{\omega_0}{Q}\right)s + \omega_0^2} \tag{4.6}$$

If $k_p = 5$, $\omega_0 = 2\pi(5000)$ rads/s, $Q = 20$, calculate the values of $|H(s)|$ for frequencies from 4500 to 5500 Hz with increments of 50 Hz.

Solution

When $s = jw$, Equation (4.6) becomes:

$$H(j\omega) = \frac{k_p[(j\omega)^2 + \omega_0^2]}{(j\omega)^2 + \left(\dfrac{\omega_0}{Q}\right)j\omega + \omega_0^2} = \frac{k_p[\omega_0^2 - \omega^2]}{\omega_0^2 - \omega^2 + j\dfrac{\omega_0}{Q}\omega} \tag{4.7}$$

MATLAB® is used to compute $H(jw)$ for various values of w.

MATLAB Script

```
% magnitude of H(j )
kp = 5;      Q = 20;    w0 = 2*pi*5000;
kj = w0/Q;
%
for i = 1:21
   w(i)  = 2*pi*(4500 + 50*(i-1));
   wh(i) = w0^2 - w(i)^2;
   h(i)  = kp*wh(i)/(wh(i) + j*kj*w(i));
   h_mag(i) = abs(h(i));        % magnitude of transfer
   function
end
% print results
for k = 1:21
   w(k)/2*pi
   h_mag(k)
end
```

The results are:

Frequency, Hz	Magnitude
4.4413e + 004	4.8654
4.4907e + 004	4.8335
4.5400e + 004	4.7898
4.5894e + 004	4.7278
4.6387e + 004	4.6363
4.6881e + 004	4.4949
4.7374e + 004	4.2643
4.7868e + 004	3.8651
4.8361e + 004	3.1428
4.8855e + 004	1.8650
4.9348e + 004	0
4.9842e + 004	1.8490
5.0335e + 004	3.1047
5.0828e + 004	3.8179
5.1322e + 004	4.2166
5.1815e + 004	4.4502
5.2309e + 004	4.5953
5.2802e + 004	4.6905
5.3296e + 004	4.7558
5.3789e + 004	4.8025
5.4283e + 004	4.8369

From the above results, the notch frequency is 493.48 KHz.

4.7 IF Statements

IF statements use relational or logical operations to determine what steps to perform in the solution of a problem. The relational operators in MATLAB® for comparing two matrices of equal size are shown in Table 4.4.

TABLE 4.4

Relational Operators

Relational Operator	Meaning
<	Less than
<=	Less than or equal
>	Greater than
>=	Greater than or equal
==	Equal
~=	Not equal

When any one of the above relational operators is used, a comparison is done between the pairs of corresponding elements. The result is a matrix of ones and zeroes, with **one** representing **TRUE** and **zero FALSE**. For example, if we have:

```
ca = [1  7  3  8  3  6];
cb = [1  2  3  4  5  6];
ca == cb
```

the answer obtained is

```
ans =
     1    0    1    0    0    1
```

The 1s indicate the elements in vectors ca and cb that are the same and 0s are the ones that are different.

There are three logical operators in MATLAB. These are shown in Table 4.5.

Logical operators work element-wise and are usually used on 0-1 matrices, such as those generated by relational operators. The & and ! operators compare two matrices of equal dimensions. If A and B are 0-1 matrices, then A&B is another 0-1 matrix with ones representing TRUE and zeroes FALSE. The NOT (~) operator is a unary operator. The expression ~C returns 1 where C is zero and 0 when C is nonzero.

There are several variations of the IF statement:

- Simple if statement
- Nested if statement
- If-else statement

- The general form of the **simple if statement** is:

    if logical expression 1

    statement group G1

    end

In the case of a simple if statement, if the logical expression 1 is true, the statement group G1 is executed. However, if the logical expression

TABLE 4.5

Logical Operators

Logical Operator Symbol	Meaning
&	and
!	or
~	not

is false, the statement group G1 is bypassed and the program control jumps to the statement that follows the end statement.

- The general form of a **nested if statement** is:

  if logical expression 1

        statement group G1

    if logical expression 2

        statement group G2

    end

        statement group G3

  end

  statement group G4

The program control is such that if expression 1 is true, then statement groups G1 and G3 are executed. If the logical expression 2 is also true, the statement groups G1 and G2 will be executed before executing statement group G3. If logical expression 1 is false, we jump to statement group G4 without executing statement groups G1, G2, and G3.

- The **if-else statement** allows one to execute one set of statements if a logical expression is true and a different set of statements if the logical statement is false. The general form of the if-else statement is:

  if logical expression 1

        statement group G1

    else

        statement group G2

  end

In the above program segment, statement group G1 is executed if logical expression 1 is true. However, if logical expression 1 is false, statement group G2 is executed.

- The **if-elseif statement** may be used to test various conditions before executing a set of statements. The general form of the if-elseif statement is:

  if logical expression 1

        statement group G1

    elseif logical expression 2

        statement group G2

> elseif logical expression 3
> > statement group G3
> elseif logical expression 4
> > statement group G4
>
> end

A statement group is executed provided the true logical expression above is true. For example, if logical expression 1 is true, then statement group G1 is executed. If logical expression 1 is false and logical expression 2 is true, then statement group G2 will be executed. If logical expressions 1, 2, and 3 are false and logical expression 4 is true, then statement group G4 will be executed. If none of the logical expressions is true, then statement groups G1, G2, G3, and G4 will not be executed. Only three elseif statements are used in the above example. More elseif statements maybe used if the application requires them. The following example illustrates the use of the if statement.

Example 4.4: Output Voltage of an Asymmetrical Limiter

The circuit shown in Figure 4.3 is a limiter. The circuit limits the output voltage to a specific value provided the input voltage exceeds or is lower than some threshold voltages. If the transfer function of the limiter is given as:

$$
\begin{aligned}
v_0(t) &= 3.0\,V & &\text{for } v_s(t) > 3.0\,V \\
&= v_s(t) & &-4.0\,V \leq v_s(t) \leq 3.0V \\
&= -4.0V & &\text{for } v_s(t) < -4.0V
\end{aligned}
\tag{4.8}
$$

where, we have assumed that the conducting diode has 0.7 V drop. Write a MATLAB® program to obtain the output voltage from 0 to 24 seconds.

Solution
MATLAB Script

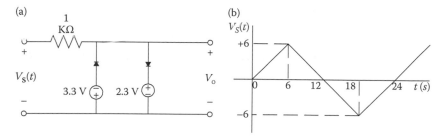

FIGURE 4.3
(a) Limiter circuit. (b) Input voltage.

```
% vo is the output voltage
% vs is the input voltage
%
%Generate the triangular wave
for    i = 1:25
       k = i-1;
     if i <= 7
         vs(i) = k;
     elseif   i >= 7 & i <= 19
         vs(i)= 12 - k;
     else
        vs(i) = -24 + k;
     end
end
% Generate output voltage using if statement
for j = 1:25
     if vs(j)>= 3.0
            vo(j) = 3.0;
     elseif vs(j) <= -4.0
            vo(j) = -4.0;
     else
            vo(j) = vs(j);
     end
end
% print results
vs
vo
```

The results are:

```
vs =

Columns 1 through 12
    0    1    2    3    4    5    6    5    4    3    2    1

Columns 13 through 24
    0   -1   -2   -3   -4   -5   -6   -5   -4   -3   -2   -1
Column 25
    0

vo =

Columns 1 through 12
    0    1    2    3    3    3    3    3    3    3    2    1

Columns 13 through 24
    0   -1   -2   -3   -4   -4   -4   -4   -4   -3   -2   -1

Column 25
    0.
```

Note that the output voltage is clipped at 3 V and –4 V.

4.8 Graph Functions

MATLAB® has built-in functions that allow one to generate x-y, polar, contour, 3-D plots, and bar charts. MATLAB also allows one to give titles to graphs, label the x- and y- axes, and add grids to graphs. In addition, there are commands for controlling the screen and scaling. Table 4.6 shows a list of MATLAB built-in graph functions. One can use MATLAB's help facility to get more information on the graph functions.

4.8.1 X-Y Plots and Annotations

The plot command generates a linear x-y plot. There are three variations of the plot command:

(a) plot(x)

(b) plot(x, y)

(c) plot(x1, y1, x2, y2, x3, y3,....,xn, yn).

TABLE 4.6

Plotting Functions

Function	Description
axis	Freezes the axis limits
bar	Plots bar chart
contour	Performs contour plots
ginput	Puts cross-hair input from mouse
grid	Adds grid to a plot
gtext	Provides mouse positioned text
histogram	Provides histogram bar graph
loglog	Does log vs. log plot
mesh	Performs 3-D mesh plot
meshdom	Provides domain for 3-D mesh plot
pause	Wait between plots
plot	Performs linear x-y plot
polar	Performs polar plot
semilogx	Does semilog x-y plot (x-logarithmic)
semilogy	Does semilog x-y plot (y-logarithmic)
stairs	Performs stair-step graph
text	Positions text at a specified location on graph
title	Used to put title on graph
xlabel	Labels x-axis
ylabel	Labels y-axis

If x is a vector, the command

plot(x)

will produces a linear plot of the elements in the vector x as a function of the index of the elements in x. MATLAB® will connect the points by straight lines. If x is a matrix, each column will be plotted as a separate curve on the same graph.

If x and y are vectors of the same length, then the command:

plot(x, y)

plots the element of x (x-axis) versus the elements of y (y-axis).

To plot multiple curves on a single graph, one can use the plot command with multiple arguments, such as:

plot(x1, y1, x2, y2, x3, y3, …, xn, yn).

The variables x1, y1, x2, y2, and so on are pairs of vectors. Each x-y pair is graphed, generating multiple lines on the plot. The above plot command allows vectors of different lengths to be displayed on the same graph. MATLAB automatically scales the plots. Also, the plot remains as the current plot until another plot is generated; in which case, the old plot is erased.

When a graph is drawn, one can add a grid, a title, a label, and x- and y-axes to the graph. The commands for grid, title, x-axis label, and y-axis label are **grid** (grid lines), **title** (graph title), **xlabel** (x-axis label), and **ylabel** (y-axis label), respectively.

To write text on a graphic screen beginning at a point (x, y) on the graphic screen, one can use the command:

text(x,y, 'text').

For example, the statement:

```
text(2.0,1.5, 'transient analysis')
```

will write the text, transient analysis, beginning at point (2.0,1.5). Multiple text commands can be used. For example, the statements:

```
plot(a1,b1,a2,b2)
text(x1,y1, 'voltage')
text(x2,y2, 'power')
```

will provide texts for two curves: a1 versus b1 and a2 versus b2. The text will be at different locations on the screen provided $x1 \neq x2$ or $y1 \neq y2$.

If the default line-types used for graphing are not satisfactory, various symbols may be selected. For example:

plot(a1, b1, '*')

draws a curve, a1 versus b1, using star(*) symbols, while

plot(a1, b1, '*', a2, b2, '+')

uses a star(*) for the first curve and the plus (+) symbol for the second curve. Other print types are shown in Table 4.7.

For systems that support color, the color of the graph may be specified using the statement:

plot(x, y, 'g')

implying, plot x versus y using green color. Line and mark style may be added to color type using the command:

plot(x, y, ' + w').

The above statement implies plot x versus y using white + marks. Other colors that can be used are shown in Table 4.8.

The argument of the plot command can be complex. If z is a complex vector, then plot(z) is equivalent to plot(real(z), imag(z)). The following example shows the use of the plot, title, xlabel, and ylabel functions.

TABLE 4.7

Print Types

Line-Types	Indicators	Point Types	Indicators
solid	-	point	.
dash	--	plus	+
dotted	:	star	*
dashdot	-.	circle	o
		xmark	x

TABLE 4.8

Symbols for Color used in Printing

Color	Symbol
red	r
green	g
blue	b
white	w
invisible	i

Example 4.5: Amplitude Modulated Wave

A block diagram of an amplitude modulation of a communication system is shown in Figure 4.4. The double-sideband suppressed carrier $s(t)$ is given as:

$$s(t) = m(t)\, c(t). \tag{4.9}$$

If

$$m(t) = 2\cos(2000\pi t)V \tag{4.10}$$

$$c(t) = 10\cos(2\pi(10^6 t))V \tag{4.11}$$

plot $s(t)$ from 0 to 60 μs.

Solution

MATLAB® Script

```
% Amplitude modulated wave
% m(t) is the message signal
% c(t) is the carrier signal
% s(t) is the modulated wave
t = 0: 0.05e-6:3.0e-6;
k = length(t)
for i = 1:k
   m(i) = 2*cos(2*pi*1000*t(i));
   c(i) = 10*cos(2*pi*1.0e + 6*t(i));
   s(i) = m(i)*c(i);
end
plot(t, s, t, s,'o')
title('Amplitude Modulated Wave')
xlabel('Time in sec')
ylabel('Voltage, V')
```

The amplitude-modulated wave is shown in Figure 4.5.

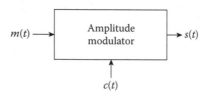

FIGURE 4.4
Block diagram of amplitude modulator.

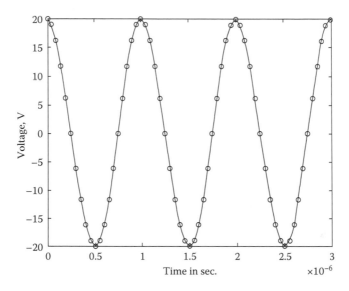

FIGURE 4.5
Amplitude modulated wave.

4.8.2 Logarithmic and Plot3 Functions

Logarithmic and semi-logarithmic plots can be generated using the commands **loglog, semilogx**, and **semilogy**. The use of the above plot commands is similar to those of the plot command discussed in the previous section. The descriptions of these commands are as follows:

loglog(x, y) – generates a plot of $\log_{10}(x)$ versus $\log_{10}(y)$

semilogx(x, y) – generates a plot of $\log_{10}(x)$ versus linear axis of y

semilogy(x, y) – generates a plot of linear axis of x versus $\log_{10}(y)$.

It should be noted that since the logarithm of negative numbers and zero do not exist, the data to be plotted on the semi-log axes or log–log axes should not contain zero or negative values.

Plot3 function can be used to do three-dimensional line plots. The function is similar to the two-dimensional **plot** function. The plot3 function supports the same line size, line style, and color options that are supported by the plot function. The simplest form of the plot3 function is:

plot(x, y, z)

where **x**, **y**, and **z** are equal-sized arrays containing the locations of the data points to be plotted.

The following example illustrates the use of the logarithmic plot.

Example 4.6: Magnitude Characteristics of a High Pass Network

The gain versus frequency of a high pass network is shown in Table 4.9. Draw a graph of the gain versus frequency.

Solution

Logarithmic scale is used for the frequency axis and linear scale for gain. The MATLAB® script is shown below.

MATLAB Script

```
% magnitude characteristics
% freq is the frequency values
freq = [30    50    100    200    500    1000    4000    6000
10000    50000    100000];

% gain is the corresponding gain
gain = [15      20   25   35   50   65   85   90   92   97   99];
% use semilog to plot gain versus frequency
semilogx(freq, gain)
title('Characteristics of a High pass Network')
xlabel('Frequency in Hz')
ylabel('Gain in dB')
```

The magnitude characteristics of the network are shown in Figure 4.6.

TABLE 4.9

Frequency versus Gain Data
of a High Pass Network

Frequency, Hz	Gain, dB
30	15
50	20
100	25
200	35
500	50
1000	65
4000	85
6000	90
10,000	92
50,000	97
100,000	99

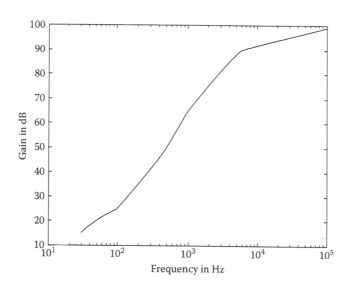

FIGURE 4.6
Gain versus frequency of a high pass network.

4.8.3 Subplot and Screen Control

MATLAB® has two display windows: a **command window** and a **graph window**. The following commands can be used to select and clear the windows:

shg	-	shows graph window
clc	-	clears command window
clg	-	clears graph window
home	-	home command cursor.

The graph window can be partitioned into multiple windows. The **subplot** command allows one to split the graph into two subdivisions or four subdivisions. Two subwindows can be arranged either top to bottom or left to right. A four-window partition will have two sub-windows on top and two subwindows on the bottom. The general form of the subplot command is:

subplot (i j k).

The digits **i** and **j** specify that the graph window is to be split into an i-by-j grid of smaller windows, arranged in i rows and j columns. The digit k specifies the kth window for the current plot. The subwindows are numbered from **left to right, top to bottom.**

For example, the command subplot (3 2 4) creates 6 subplots in the current figure and makes subplot 4, the current plotting window. This is shown in Table 4.10.

The following example illustrates the use of the subplot command.

TABLE 4.10

Numbering of Subwindows for
Subplot Command, Subplot(3, 2, 4)

1	2
3	4 (current figure)
5	6

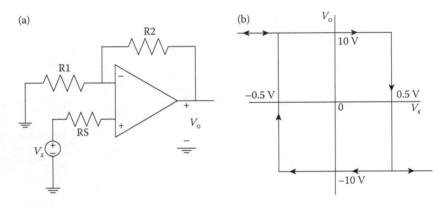

FIGURE 4.7
(a) Schmitt trigger circuit. (b) Its transfer characteristics.

Example 4.7: Input and Output Voltages of a Schmitt Trigger Circuit

For the inverting Schmitt trigger circuit shown in Figure 4.7a, $R1 = 1$ KΩ, $R2 = 19$ KΩ, and $RS = 2$ KΩ. The transfer characteristics of the circuit are shown in Figure 4.7b. If the input voltage, $v_s(t)$ is a noisy signal given as:

$$v_s(t) = 1.5\sin(2\pi f_0 t) + 0.8n(t) \tag{4.12}$$

where

$$f_0 = 500 \text{ Hz}$$

and

$$n(t) \text{ is a normally distributed white noise,}$$

write a MATLAB® program to find the output voltage. Plot both the input and output waveforms of the Schmitt trigger.

Solution

If $V_0(t)$ is the output voltage at time t, then the input and output voltages are related by the expressions:

$$v_0(t) = -10\,V \quad \text{if} \quad v_s(t) \geq 0.5\,V$$

$$= -10\,V \quad \text{if} \quad -0.5\,V < v_s(t) < 0.5\,V \quad \text{and} \quad v_0(t-1) = -10\,V \quad (4.13)$$

$$= +10\,V \quad \text{if} \quad -0.5\,V < v_s(t) < 0.5\,V \quad \text{and} \quad v(_0t-1) = +10\,V$$

$$= +10\,V \quad \text{if} \quad v_s(t) \leq -0.5\,V$$

If statements will be used to execute relationships shown in Equation 4.13.

MATLAB Script

```
% vo is the output voltage
% vs is the input voltage
% Generate the sine voltage
t = 0.0:0.1e-4:5e-3;
fo = 500;     % frequency of sine wave
len = length(t)
for i = 1:len
    s(i) = 1.5*sin(2*pi*fo*t(i));
    % Generate a normally distributed white noise
    n(i) = 0.8*randn(1);
    % generate the noisy signal
    vs(i) = s(i) + n(i);
end
% calculation of output voltage
%
len1 = len -1;
for i = 1:len1
   if    vs(i + 1) >= 0.5;
      vo(i + 1) = -10;
   elseif vs(i + 1) > -0.5 & vs(i + 1) < 0.5 & vo(i) ==
   -10
         vo(i + 1) = -10;
   elseif vs(i + 1) > -0.5 & vs(i + 1) < 0.5 & vo(i) ==
   +10
         vo(i + 1) = 10;
      else
         vo(i + 1) = +10;
      end
end
%
% Use subplots to plot vs and vo
      subplot (211), plot (t(1:40), vs(1:40))
      title ('Noisy time domain signal')
      subplot (212), plot (t(1:40), vo(1:40))
      title ('Output Voltage')
      xlabel ('Time in sec')
```

The input and output voltages are shown in Figure 4.8.

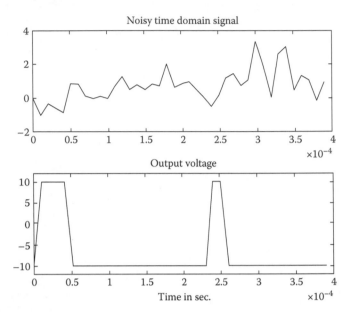

FIGURE 4.8
Input and output voltages.

4.8.4 Bar Plots

The bar function is used to plot bar plots. The general form of the command is:

bar(x, y) - creates a vertical bar plot, where the values in x are used to label each bar and the values in y are used to determine the height of the bar. There are other variations of the bar function, such as:

barh(x, y) - This function creates a horizontal bar plot. The values in x are used to label each bar and the values in y are used to determine the horizontal length of the bar.

bar3(x, y) - This function renders bar charts to have three dimensional appearances.

bar3h(x, y) - This function is similar to barh(x, y), but it gives the bar chart a three dimensional appearance.

4.8.5 Hist Function

The hist function can be used to calculate and plot the histogram of a set of data. Histogram shows the distribution of a set of values in a data set. The general format for using the function is as follows:

hist(x) - calculates and plots the histogram of values in a data set x by using 10 bins.

hist(x, n) – calculates and plots the histogram of values in a data set x by using n equally spaced bins.

Hist(x, y) – calculates and plots the histogram of values of x using the bins with centers specified by the values of the vector y.

Example 4.8: Plot of a Gaussian Random Data

Plot the histogram of a Gaussian distributed data with mean of 5.0 and standard deviation of 1.5.

Solution

A random data with 5000 points, Gaussian distributed, with mean value of zero and standard deviation 2 can be generated with the equation:

$$R\_data = 1.5*randn(500, 1) + 5.0.$$

Plot the histogram of the random Gaussian data using 20 evenly spaced bins.

Solution

MATLAB® Script

```
% Generate the random Gaussian data
r_data = 1.5*randn(5000,1) + 5.0;
hist(r_data, 20)
title('Histogram of Gaussian Data')
```

The histogram of the Gaussian random data is shown in Figure 4.9.

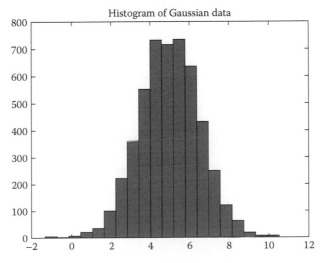

FIGURE 4.9
Histogram of Gaussian data.

4.8.6 Stem Plots

The **stem** function generates a point plot with lines or stems, connecting the point to the x-axis. It is normally used to plot discrete sequence data. Its usage is:

Stem(z) - creates a plot of data points in vector z connected to the horizontal axis. An optional character string can be used to specify line style.

Stem(x, z) - plots the data points in z at values specified in x.

Example 4.9: Convolution Between Two Discrete Data

The convolution between two discrete data X and Y is given as Z = conv(X, Y). If X = [0 1 2 3 4 3 2 1 0] and Y = [0 1 2 1 0], find Z. Plot X, Y, and Z.

Solution

MATLAB® Script

```
% The convolution function Z = conv(X, Y) is used
X = [0 1 2 3 4 3 2 1 0];
Y = [0 1 2 1 0];
Z = conv(X, Y);
stem(Z), title('Convolution Between X and Y')
```

The plots are shown in Figure 4.10.

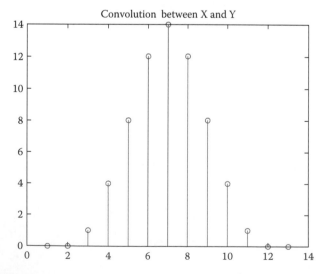

FIGURE 4.10
Convolution between X and Y.

4.9 Input/Output Commands

MATLAB® has commands for inputting information in the command window and outputting data. Examples of input/output commands are **echo, input, pause, keyboard, break, error, display, format, and fprintf.** Brief descriptions of these commands are shown in Table 4.11.

Break

The **break** command may be used to terminate the execution of **for** loop. If the break command exits in an innermost part of a nested loop, the break command will exit from that loop only. The break command is useful in exiting a loop when an error condition is detected.

Disp

The **disp** command displays a matrix without printing its name. It can also be used to display a text string. The general form of the **disp** command is:

disp(x)
disp('text string')

disp(x) will display the matrix. Another way of displaying matrix x is to type its name. This is not always desirable since the display will start with a leading "x = ". Disp ('text string') will display the text string in quotes. For example, the MATLAB statement:

disp('3-by-3 identity matrix')

will result in

3-by-3 identity matrix.

TABLE 4.11

Some Input/Output Commands

Command	Description
break	Exits while or for loops
disp	Displays text or matrix
echo	Displays m-files during execution
error	Displays error messages
format	Displays output display to a particular format
fprintf	Displays text and matrices and specifies format for printing values
input	Allows user input
pause	Causes an m-file to stop executing. Pressing any key causes interruptions of program execution.

Echo

The **echo** command can be used for debugging purposes. The echo command allows commands to be viewed as they execute. The echo can be enabled or disabled.

echo on - enables the echoing of commands
echo off - disables the echoing of commands
echo - by itself toggles the echo statement

Error

The error command curve causes an error return from the m-files (discussed in chapter 4) to the keyboard and displays a user written message. The general form of the command is:

Error('messages for display').

Consider the following MATLAB statements:

x = input('Enter age of student');
if x < 0
        error('wrong age was entered, try again')
end
x = input('Enter age of student')

Format

The format controls the format of an output. Table 4.12 shows some formats available in MATLAB. By default, MATLAB, displays numbers in "short" format (five significant digits). Format compact suppresses linefeed that appear between matrix displays, thus allowing more lines of information to be seen on the screen. **Format loose** reverts to the less compact display. Format compact and format loose do not affect the numeric format.

TABLE 4.12

Format Displays

Command	Meaning
format short	five significant decimal digits
format long	15 significant digits
format short e	Scientific notation with 5 significant digits
format long e	Scientific notation with 15 significant digits
format hex	Hexadecimal
format +	+ printed if value is positive, − if negative; space is skipped if value is zero

fprintf

The **fprintf** can be used to print both text and matrix values. The format for printing the matrix can be specified, and line feed can also be specified. The general form of this command is:

fprintf('text with format specification,' matrices).

For example, the following statements:

```
res = 1.0e+6;
fprintf('The value of resistance is %7.3e Ohms\n', res)
```

when executed will yield the output

```
The value of resistance is 1.000e+006 Ohms
```

The format specifier **%7.3e** is used to show where the matrix value should be printed in the text. 7.3e indicates that the resistance value should be printed with an exponential notation of seven digits, three of which should be decimal digits. Other format specifiers are:

%c – Single character
%d – Decimal notation (signed)
%e – Exponential notation (using a lowercase e as in 2.051e + 01)
%f – Fixed-point notation
%g – signed decimal number in either %e or %f format, whatever is shorter.

The text with format specification should end with **\n** to indicate the end of line. However, we can also use **\n** to get line feeds as represented by the following example:

```
r1 = 1500;
fprintf('resistance is \n%f Ohms \n',r1)
```

the output is

```
resistance is
1500.000000 Ohms
```

Input

The input command displays a user-written text string on the screen, waits for an input from the keyboard, and assigns the number entered on the keyboard as the value of a variable. If the user enters a single number, it may be typed in. However, if the user enters an array, it must be enclosed in brackets. In either case, whatever is typed in will be stored in a variable. For example, if one types the command:

```
r - input('Please enter the three resistor values');
```

when the above command is executed, the text string 'Please, enter the three resistor values' will be displayed on the terminal screen. The user can then type an expression such as:

```
[12    14    9]
```

The variable r will be assigned a vector [12 14 9]. If the user strikes the return key, without entering an input, an empty matrix will be assigned to r.

To return a string typed by a user as a text variable, the input command may take the form:

```
x = Input('Enter string for prompt', 's')
```

For example, the command:

```
x = input('What is the title of your graph', 's')
```

when executed, will echo on the screen, 'What is the title of your graph.' The user can enter a string such as 'Voltage (mV) versus Current (mA).' One will get the echo:

```
x =
'Voltage (mV) versus Current (mA).'
```

Pause

The **pause** command stops the execution of m-files. The execution of the m-file resumes upon pressing any key. The general forms of the pause command are:

pause

pause(n)

Pause stops the execution of m-files until a key is pressed. **Pause(n)** stops the execution of m-files for n seconds before continuing. The pause command can be used to stop m-files temporarily when plotting commands are encountered during program execution. If pause is not used, the graphics are momentarily visible.

The following example uses the MATLAB input, fprint, and disp commands.

Example 4.10: Equivalent Resistance of Series Connected Resistor

Write a MATLAB® program that will accept values of resistors connected in series and find the equivalent resistance. The values of the resistors will be entered from the keyboard.

Solution

We shall use the MATLAB input command to accept the input of the elements, the fprintf command to output the result, and disp command to display the text string.

MATLAB Script

```
% input values of the resistors in input order
%
%
disp('Enter resistor values with spaces between them and
enclosed in brackets')
res = input('Enter resistor values')
num = length(res);   % number of elements in array res
requiv = 0;
  for i = 1:num
    requiv = requiv + res(i);
end
%
fprintf('The Equivalent Resistance is %8.3e Ohms',
requiv)
```

If you enter the values [2   3   7   9], you get the result:

```
res =
     2   3   7   9
The Equivalent Resistance is 2.100e + 001 Ohms
```

Problems

4.1   Find the nodal voltages V1, V2, V3, and V4 of Figure P4.1. The resistances are in Ohms.

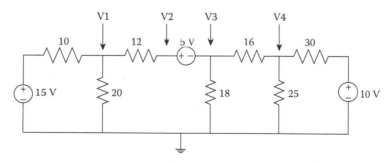

FIGURE P4.1
Circuit for Problem 4.1.

4.2   For the network shown in Figure P4.2, find nodal voltages, V1, V2, V3, and V4. The resistances are in Ohms.

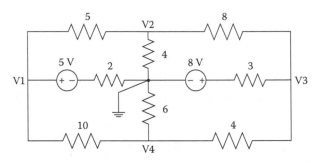

FIGURE P4.2
Circuit for Problem 4.2.

4.3   Use loop analysis to find the current IO. The resistances are in Ohms.

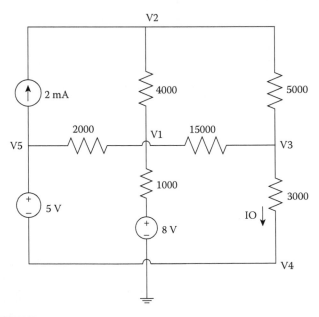

FIGURE P4.3
Circuit for Problem 4.3.

4.4   Find the loop currents I_1, I_2, and I_3, for the ladder network. The resistances are in Ohms.

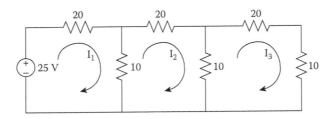

FIGURE P4.4
Ladder network.

4.5   Simplify the following complex numbers and express them in rectangular and polar form.

(a) $za = 18 + j12 + \dfrac{(20 + j40)(5 - j15)}{25 + j25}$

(b) $zb = \dfrac{10(-5 + j13)(4 + j4)}{(1 + j2)(2 + j5)(-5 + j3)}$

(c) $zc = 0.2 + j7 + 4.7e^{j0.5} + (2 + j3)e^{-j0.6\pi}$

4.6   Find the input impedance of the circuit shown in Figure P4.6. The impedances are in Ohms.

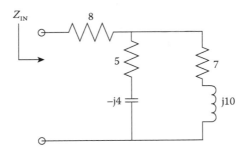

FIGURE P4.6
Circuit for Problem 4.6.

4.7   The closed loop gain, G, of an operational amplifier with finite open loop gain of A is given as:

$$G = \dfrac{-\dfrac{R_2}{R_1}}{1 + \left(\dfrac{1 + \dfrac{R_2}{R_2}}{A} \right)}$$

If $R_2 = 20$ KΩ and $R_2 = 1$ KΩ, find the closed loop gain for the following values of the open loop gain : 10^2, 10^3, 10^4, 10^5, 10^6, and 10^7.

4.8   For the Figure P4.8, find the equivalent admittance (in polar form) for the following frequencies: 1 KHz, 4 KHz, 7 KHz, and 10 KHz.

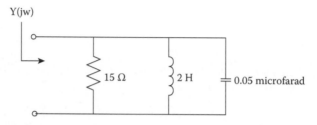

FIGURE P4.8
Parallel RLC circuit.

4.9   A limiter circuit shown in Figure P4.9. Assuming that a conducting diode has 0.7 V drop, the relation between $i_s(t)$ and $v_s(t)$ is given as:

$$i_s(t) = 0 \qquad\qquad \text{for } -6\text{V} < v_s(t) < 3.0\text{V}$$

$$= \frac{(v_s(t)-3)}{2000} \qquad \text{for } v_s(t) > 3.0\text{V}$$

$$= \frac{(v_s(t)+6.0)}{1000} \qquad \text{for } v_s(t) < -6.0\text{V}.$$

If $v_s(t)$ is a square wave with peak value of 10 V, average value of 0 V and period of 4 ms, plot the input current $i_s(t)$ for one period of the input voltage. The resistances are in Ohms.

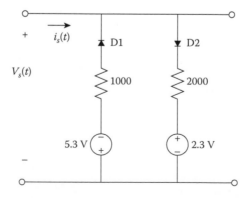

FIGURE P4.9
Limiter circuit.

4.10 The current following through the drain of a MOSFET is given:

$$i_{DS} = k_p(V_{GS} - V_T)^2 \text{ A}.$$

If $V_T = 0.8$ V and $k_p = 4$ mA/V², plot i_{DS} for the following values of V_{GS}: 2, 2.5, and 3 V.

4.11 The equivalent impedance of the circuit, shown in Figure P4.11, is given as:

$$z(j\omega) = R + \frac{jwL}{1 - w^2 LC}.$$

If $L = 1$ mH, $C = 10$ μF, and $R = 100$ Ω, plot the magnitude of the input impedance for $w = 10$, 100, 1000, 1.0e04, and 1.0e05 rads/s.

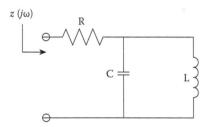

FIGURE P4.11
Circuit for Problem 4.11.

4.12 The equivalent impedance of a circuit given as:

$$z(j\omega) = R + jwL - \frac{j}{wC}.$$

If $L = 1$ mH, $C = 0.01$ μF, and $R = 10$ Ω, plot the magnitude of the input impedance.

4.13 Use the stem function to plot the convolution between the two discrete data x and y, if:
(a) X = [0 1] and Y = [0 1 1 0 1 1 0]
(b) X = [1 2 4 6 8 6 4 2 1] and Y = [1 0 1]
(c) X = [0 −1 0 1 0 −1 0 1] and Y = [2 0 1 0 −1 0 −2 0].

4.14 Determine the loop currents shown in Figure P4.14.

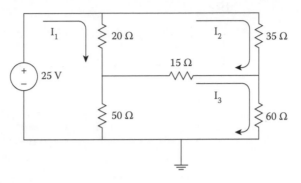

FIGURE P4.14
Circuit for Exercise 4.14.

4.15 Find the power dissipated by the 10 Ω resistor and the voltage V_0.

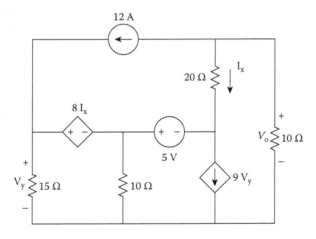

FIGURE P4.15
Circuit for Exercise 4.15.

4.16 The voltage v_D and i_D of a diode is related by the expression:

$$i_D = I_s \exp\left(\frac{v_D}{nV_T}\right).$$

If $i_D = 10^{-16}$, $n = 1.5$, and $V_T = 26$ mV, plot i_D for diode voltages ranging from 0 to 0.65 V.

Bibliography

1. Attia, J. O. *Electronics and Circuit Analysis Using MATLAB®*. 2nd ed. Boca Raton, FL: CRC Press, 2004.
2. Biran, A., and M. Breiner. *MATLAB® for Engineers*. White Plains, NY: Addison-Wesley, 1995.
3. Boyd, Robert R. *Tolerance Analysis of Electronic Circuits Using MATLAB®*. Boca Raton, FL: CRC Press, 1999.
4. Chapman, S. J. *MATLAB® Programming for Engineers*. Tampa, FL: Thompson, 2005.
5. Davis, Timothy A., and K. Sigmor. *MATLAB® Primer*. Boca Raton, FL: Chapman & Hall/CRC, 2005.
6. Etter, D. M. *Engineering Problem Solving with MATLAB®*. 2nd ed. Upper Saddle River, NJ: Prentice Hall, 1997.
7. Etter, D. M., D. C. Kuncicky, and D. Hull. *Introduction to MATLAB® 6*. Upper Saddle River, NJ: Prentice Hall, 2002.
8. Gilat, Amos. *MATLAB®, An Introduction With Applications*. 2nd ed. New York: John Wiley & Sons, 2005.
9. Gottling, J. G. *Matrix Analysis of Circuits Using MATLAB®*. Upper Saddle River, NJ: Prentice Hall, 1995.
10. Hahn, Brian D., and Daniel T. Valentine. *Essential MATLAB® for Engineers and Scientists*. 3rd ed. New York and London: Elsevier, 2007.
11. Herniter, Marc E. *Programming in MATLAB®*. Florence, KY: Brooks/Cole Thompson Learning, 2001.
12. Howe, Roger T., and Charles G. Sodini. *Microelectronics, An Integrated Approach*. Upper Saddle River, NJ: Prentice Hall, 1997.
13. Moore, Holly. *MATLAB® for Engineers*. Upper Saddle River, NJ: Pearson Prentice Hall, 2007.
14. *Using MATLAB®, The Language of Technical Computing, Computation, Visualization, Programming, Version 6*. Natick, MA: MathWorks, Inc. 2000.

5

MATLAB® Functions

In this chapter, MATLAB® functions that will allow the user to process data from PSPICE simulations will be discussed. The chapter begins with a discussion of m-files (script and function file). Some built-in MATLAB mathematical and statistical functions will be introduced. In addition, four MATLAB functions, diff, quad, quad8, and fzero, will be discussed. Furthermore, input/output functions will also be covered. The chapter ends with methods of accessing results of PSPICE simulations by MATLAB.

5.1 M-Files

Normally, when single line commands are entered, MATLAB® processes the commands immediately and displays the results. MATLAB is also capable of processing a sequence of commands that are stored in files with extension **m.** MATLAB files with extension m are called **m-files**. The latter are ASCII text files and are created with a text editor or word processor.

To list m-files in the current directory on your disk, you can use the MATLAB command **what**. The MATLAB command, **type**, can be used to show the contents of a specified file.

M-files can either be script or function. Script and function files contain a sequence of commands. However, function files take arguments and return values.

5.1.1 Script Files

Script files are especially useful for analysis and design problems that require long sequences of MATLAB® commands. With script file written using a text editor or word processor, the file can be invoked by entering the name of the m-file, without the extension. Statements in a script file operate globally on the workspace data.

Normally, when m-files are executing, the commands are not displayed on screen. The MATLAB echo command can be used to view m-files while they are executing. The MATLAB programs in Examples 3.1 through 3.8 are script files.

5.1.2 Function Files

Function files are m-files that are used to create new MATLAB® functions. Variables defined and manipulated inside a function file are local to the function, and they do not operate globally on the workspace. However, arguments may be passed into and out of a function file.

The general form of a function file is:

function variable(s) = function_name (arguments)
% help text in the usage of the function
%
.
.

The following is a summary of the rules for writing MATLAB m-file functions:

1. The word function appears as the first word in a function file. This is followed by an output argument, an equal sign and the function name. The arguments to the function follow the function name and are enclosed within parentheses.

2. The information that follows the function, beginning with the % sign, shows how the function is used and what arguments are passed. This information is displayed if help is requested for the function name.

3. MATLAB can accept multiple input arguments and multiple output arguments can be returned.

4. If a function is going to return more than one value, all the values should be returned as a vector in the function statement. For example,

$$\text{function[mean, variance]} = \text{data\_in}(x)$$

will return the mean and variance of a vector x. The mean and variance are computed with the function.

5. If a function has multiple input arguments, the function statement must list the input arguments. For example,

$$\text{function[mean, variance]} = \text{data}(x, n).$$

The following example illustrates the usage of the m-file.

Example 5.1· Equivalent Resistance of Parallel-Connected Resistors

Write a function to find the equivalent resistance of parallel-connected resistors, R1, R2, R3,…, Rn.

Solution

MATLAB® Script

```
function req = equiv_pr(r)
% equiv_pr is a function program for obtaining
%       the equivalent resistance of series
%       connected resistors
% usage: req = equiv_pr(r)
%       r is an input vector of length n
%
n = length(r);  % number of resistors
tmp = 0.0;
for i = 1:n
  tmp = tmp + 1/r(i);
end
req = 1/tmp;
```

The above MATLAB script can be found in the function file equiv_pr.m.

Suppose we want to find the equivalent resistance of the parallel-connected resistors 2, 6, 7, 9, and 12 ohms. The following statements can be typed in the MATLAB command window to reference the function equiv_pr:

```
a = [2 6 7 9 12];
Rparall = equiv_pr(a)
```

The result obtained from MATLAB is:

```
Rparall =
        0.9960
```

The equivalent resistance is 0.996 Ohms.

5.2 Mathematical Functions

A partial list of mathematical functions that are available in MATLAB® is shown in Table 5.1. A brief description of the various functions is also given below.

TABLE 5.1

Common Mathematical Functions

Function Name	Explanation of Function
abs(x)	Absolute value or magnitude of complex number. Calculates $|x|$
acos(x)	Inverse cosine; $\cos^{-1}(x)$, the results are in radians
angle(x)	Four-quadrant angle of a complex number, phase angle of complex number x in radians
asin(x)	Inverse sine, calculates $\sin^{-1}(x)$ with results in radians
atan(x)	Calculates $\tan^{-1}(x)$, with the results in radians
atan2(x,y)	Four-quadrant inverse. This function calculates $\tan^{-1}(y/x)$ over all four quadrants of the circle. The result is in radians in the range $-\pi$ to $+\pi$.
ceil(x)	Round x to the nearest integer toward positive infinity, thus ceil(4.2) = 4; ceil(−3.3) = −3
conj(x)	Complex conjugate, i.e., x = 3 + j7; conj(x) = 3 − j7
cos(x)	Calculates cosine of x, with x in radians
exp(x)	Exponential, i.e., it calculates e^x
fix(x)	Round x to the nearest integer toward zero; fix(4.2) = 4, fix(3.3) = 3
floor(x)	Rounds x to the nearest integer towards minus infinity, floor(4.2) = 4 and floor(3.3) = 3
imag(x)	Complex imaginary part of x
log(x)	Natural logarithm: $\log_e(x)$
log10(x)	Common logarithm: $\log_{10}(x)$
real(x)	Real part of complex number x
rem(x,y)	Remainder after division of (x/y)
round(x)	Round toward nearest integer
sin(x)	Sine of x, with x in radians
sqrt(x)	Square root of x
tan(x)	Tangent of x

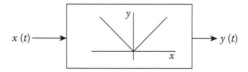

FIGURE 5.1
Block diagram of absolute value circuit.

Example 5.2: Generation of a Full-Wave Rectifier Waveform

A full-wave rectifier waveform can be generated by passing a signal through an absolute value detector, whose block diagram is shown in Figure 5.1. If the input signal is $x(t) = 10\sin(120\pi t)$, and $y(t) = |x(t)|$, write a MATLAB® program to plot $x(t)$ and $y(t)$.

Solution

MATLAB Script

```
% x(t) in the input
% y is the output
period = 1/60;
period2 = 2*period;
inc = period/100;
npts = period2/inc;
for i = 1:npts
    t(i) = (i-1)*inc;
    x(i) = 10*sin(120*pi*t(i));
    y(i) = abs(x(i));
end
% plot x and y
subplot(211), plot(t,x)
ylabel('Voltage,V')
title('Input signal x(t)')
subplot(212), plot(t,y)
ylabel('Voltage,V')
xlabel('Time in seconds')
title('Output Signal y(t)')
```

The plots are shown in Figure 5.2. From Figure 5.2, the output signal is a rectified version of the input waveform.

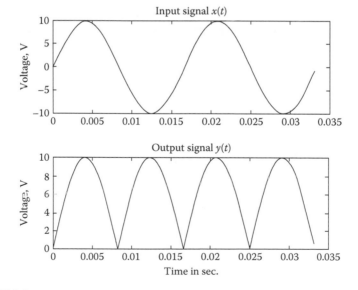

FIGURE 5.2
Input sine waveform $x(t)$ and output waveform $y(t)$.

5.3 Data Analysis Functions

In MATLAB®, data analyses are performed on column-oriented matrices. Different variables are stored in the individual column cells, and each row represents different observations of each variable. A data consisting of 10 samples in 4 variables would be stored in a matrix of size 10-by-4. Functions act on the elements in the column. Table 5.2 gives a brief description of various MATLAB functions for performing data analysis.

TABLE 5.2

Data Analysis Functions

Function	Description
corrcoef(x)	Determines correlation coefficients
cov(x)	Obtains covariance matrix
cross(x, y)	Determines cross product of vectors x and y
cumprod(x)	Finds a vector of the same size as x containing the cumulative products of the values from x. If x is a matrix, cumprod(x) returns a matrix the same size as x containing cumulative products of values from the columns of x.
cumsum(x)	Obtains a vector of the same size as x containing the cumulative sums of values from x. If x is a matrix, cumsum(x) returns a matrix the same size as x containing cumulative values from the columns of x.
diff(x)	Computes the differences between elements of an array x. It approximates derivatives. The diff function is discussed in detail in Section 5.3.
dot(x, y)	Determines the dot product of vectors x and y
hist(x)	Draws the histogram or the bar chart of x
max(x)	Obtains the largest value of x. If x is a matrix, max(x) returns a row vector containing the maximum elements of each column.
[y, k] = max(x)	Obtains the maximum value of x and the corresponding locations (indices) of the first maximum value for each column of x.
mean(x)	Determines the mean or the average value of the elements in the vector. If x is a matrix, mean(x) returns a row vector that contains the mean value of each column.
median(x)	Finds the median value of the elements in the vector x. If x is a matrix, this function returns a row vector containing the median value of each column.
min(x)	Finds the smallest value of x. If x is a matrix, min(x) returns a row vector containing the minimum values from each column.
[y, k] = min(x)	Obtains the smallest value of x and the corresponding locations (indices) the first minimum value from each column of x.
prod(x)	Calculates the product of the elements of x. If x is a matrix, prod(x) returns a row vector that contains the product of each column.
rand(n)	Generates random numbers. If n = 1, a single random number is returned. If n > 1, n-by-n matrix of random numbers are generated, rand(n) generates the random numbers uniformly distributed in the interval [0,1]
rand(m, n)	Generates an m-by-n matrix containing uniformly distributed random numbers between 0 and 1.

TABLE 5.2

Data Analysis Functions (Continued)

Function	Description
rand('seed', n)	Sets the seed number of the random number generator to n: If rand is called repeatedly with the same seed number, the sequence of random numbers become the same.
rand('seed')	Returns the current value of the "seed" values of the random number generator.
rand(m,n)	Generates an m-by-n matrix containing random numbers uniformly distributed between 0 and 1.
randn(n)	Produces an n-by-n matrix containing normally distributed (Gaussian) random numbers with a mean of zero and variance of 1.
randn(m,n)	Produces an m-by-n matrix containing normally distributed (Gaussian) random numbers with a mean of zero and variance of 1. To convert Gaussian random number r_n with mean value of zero and variance of 1 to a new Gaussian random number with mean of μ and standard deviation σ, we use the conversion formula: $$X = \sigma \cdot r_n + \mu.$$ Thus, random number data with 200 values, Gaussian distributed with mean value of 4 and standard deviation of 2 can be generated with the equation: $$\text{data\_g} = 2.\text{randn}(1,200) + 4.$$
sort(x)	Sort the rows of a matrix a in ascending order
std(x)	Calculates and returns the standard deviation of x if it is a one dimensional array. If x is a matrix a row vector containing the standard deviation of each column is computed and returned.
sum(x)	Calculates and returns the sum of the elements in x. If x is a matrix, this function calculates and returns a row vector that contains the sum of each column
trapz(x,y)	Trapezoidal integration of the function y = f(x). A detailed discussion of this function is in Section 5.5.

Example 5.3: Statistics of Resistors

Resistances of 3 bins containing 1 KΩ, 10 KΩ, and 50 KΩ resistors, respectively, were measured using a multimeter. Ten resistors selected from the three bins have values shown in Table 5.3. For each resistor bin, determine the mean, median, and standard deviation.

Solution

The data shown in Table 5.3 is stored as a 3-by-10 matrix y.

MATLAB® Script

```
% This program computes the mean, median, and standard
% deviation of resistors in bins
% the data is stored in matrix y
y = [1050 10250 50211;
```

```
992     9850    52500;
1021    9850    52500;
980     9752    53700;
1070    10102   48800;
940     9920    51650;
1005    10711   49220;
998     9520    54170;
1021    10550   46840;
987     9870    51100];
%
% Calculate the mean
mean_r = mean(y);
% Calculate the median
median_r = median(y);
% Calculate the standard deviation
std_r = std(y);
% Print out the results
fprintf('Statistics of Resistor Bins\n\n')
fprintf('Mean of 1K,10K, 50K bins,respectively:%7.3e,
%7.3e, %7.3e \n', mean_r)
fprintf('Median of 1K, 10K, 50K bins: respectively :%7.3e,
%7.3e, %7.3e\n', median_r)
fprintf('Standard Deviation of 1K, 10K, 50K bins,
respectively:%7.8e, %7.8e , %7.8e \n', std_r)
```

The results are:

```
Statistics of Resistor Bins
Mean of 1 K,10 K, 50 K bins, respectively: 1.006e + 003,
1.004e + 004, 5.107e + 004;
Median of 1 K, 10 K, 50 K bins, respectively: 1.002e + 003,
9.895e + 003, 5.138e + 004;
Standard Deviation of 1 K, 10 K, 50 K bins, respectively:
3.67187509e + 001, 3.69243446e + 002, 2.31325103e + 003.
```

TABLE 5.3

Resistances in 1 KΩ, 10 KΩ, and 50 KΩ Resistor Bins

Number	1 KΩ Resistor Bin	10 KΩ Resistor Bin	50 KΩ Resistor Bin
1	1050	10,250	50,211
2	992	9850	52,500
3	1021	10,460	47,270
4	980	9752	53,700
5	1070	10,102	48,800
6	940	9920	51,650
7	1005	10,711	49,220
8	998	9520	54,170
9	1021	10,550	46,840
10	987	9870	51,100

Before doing the next example, let us discuss the MATLAB function **freqs.** The latter function is used to obtain the frequency response of a transfer function. The general form is:

$$H(s) = \text{freqs(num, den, range)} \tag{5.1}$$

where

$$H(s) = \frac{b_m s^m + b_{m-1} s^{m-1} + \ldots + b_1 s + b_0}{a_n s^n + a_{n-1} s^{n-1} + \ldots + a_1 s + a_0} \tag{5.2}$$

$$\text{num} = [b_m \ b_{m-1} \ \ldots \ b_1 \ b_0] \tag{5.3}$$

$$\text{den} = [a_n \ a_{n-1} \ \ldots \ a_1 \ a_0] \tag{5.4}$$

range is a range of frequencies; and
hs is the frequency response (in complex form).

Example 5.4: Center Frequency of Band-Reject Filter

The transfer function of a band-reject filter is given as:

$$H(s) = \frac{s^2 + 9.859 \times 10^8}{s^2 + 3140s + 9.859 \times 10^8} \tag{5.5}$$

Find the center frequency.

Solution
MATLAB® Script

```
% numerator and denominator polynomial
num = [1  0  9.859e8];
den = [1  3.14e3  9.859e8];
w = logspace(-3,5,5000);
hs = freqs(num, den, w);      % finds frequency
f = w/(2*pi);   %finds frequency from rad/s to Hz
mag = 20*log10(abs(hs));      %magnitude of hs
% find minimum value of magnitude and its index
[mag_m floc] = min(mag);
% minimum frequency
fmin = f(floc);
%print results
fprintf('Minimum Magnitude (dB) is %8.4e\n', mag_m)
fprintf('Minimum frequency is %8.4e\n', fmin)
plot(f,mag)
```

The results obtained are:

Minimum Magnitude (dB) is −3.1424e + 001
Minimum frequency is 5.0040e + 003.

5.4 Derivative Function (diff)

If f is a row or column vector

$$f = [f(1)\, f(2) \ldots f(n)] \qquad (5.6)$$

then the **diff(f)** function returns a vector containing the difference between adjacent elements, i.e.:

$$\text{diff}(f) = [f(2)-f(1), f(3)-f(2) \ldots f(n)-f(n-1)]. \qquad (5.7)$$

The output vector diff(f) will be one element less than the input vector f.

Numerical differentiation can be obtained using the backward difference expression:

$$f'(x_n) = \frac{f(x_n) - f(x_{n-1})}{x_n - x_{n-1}} \qquad (5.8)$$

or by the forward difference

$$f'(x_n) = \frac{f(x_{n+1}) - f(x_n)}{x_{n+1} - x_n}. \qquad (5.9)$$

The derivative of $f(x)$ can be obtained by using the MATLAB® **diff** function as:

$$f'(x) \cong \frac{\text{diff}(f)}{\text{diff}(x)}. \qquad (5.10)$$

We use the MATLAB function **diff** in the following example.

Example 5.5: Differentiator Circuit with Noisy Input Signal

An operational amplifier differentiator has input and output voltages related by the expression:

$$V_0(t) = -k\frac{d}{dt}V_{IN}(t), \quad k = 0.0001. \qquad (5.11)$$

If the input voltage is given as:

$$V_{IN}(t) = \sin(2\pi f_0 t) + 0.2n(t)$$

where
  $f_0 = 500$ HZ; and
  $n(t)$ is a normally distributed white noise.
  Sketch $V_0(t)$ and $V_{in}(t)$ using the subplot command.

Solution

MATLAB® Script

```
% Differentiator circuit with noisy input
%
% generate input signal
%
t = 0.0:5e-5:6e-3;
k = -0.0001;
f0 = 500;
m = length(t);
% generate sine wave portion of signal
for i = 1:m
  s(i) = sin(2*pi*f0*t(i));
  % generate a normally distributed white noise
  n(i) = 0.2*randn(1);
  % generate noisy signal
  vin(i) = s(i) + n(i);
end
Subplot(211), plot(t(1:100), vin(1:100))
Title ('Noisy Input Signal')
% derivative of input signal is calculated using
% backward difference
dvin = diff(vin)./diff(t);
% output voltage is calculated
vout = k* dvin;
% plot the output voltage
subplot(212), plot(t(2:101), vout(1:100))
title('Output Voltage of Differentiator')
xlabel('Time in s')
```

Figure 5.3 shows the input and output of a differentiator.

5.5 Integration Function (quad, quad8, trapz)

The **quad** function uses an adaptive recursive Simpson's rule. However, the **quad8** function uses an adaptive recursive Newton Cutes 8 panel rule. The **quad8** function is better than **quad** at handling functions with "soft" singularities such as $\int \sqrt{x}dx$. Suppose we want to find S given as:

$$S = \int_a^b funct(x)dx. \qquad (5.12)$$

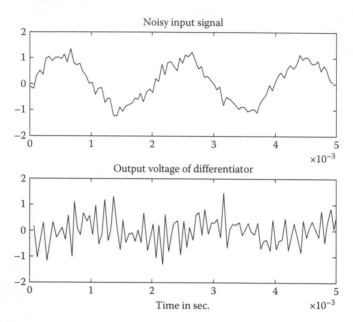

FIGURE 5.3
Input and output voltages of a differentiator.

The general form of **quad** and **quad8** functions that can be used to find S is:

$$quad('funct', a, b, tol, trace)$$

$$quad8('funct', a, b, tol, trace)$$

where
   funct is a MATLAB® function name (in quotes) that returns a vector of values $f(x)$ for a given vector of input values x;
   a is the lower limit of integration;
   b is the upper limit of integration;
   tol (optional) is the tolerance limit set for stopping the iteration of the numerical integration. The iteration continues until the relative error is less than tol. The default value is 1.0e–3; and
   trace (optional) allows the plot of a graph showing the process of numerical integration. If trace is nonzero, a graph is plotted. The default value is zero.

The **quad** and **quad8** functions use an argument that is an analytic expression of the integrand. This facility allows the functions (quad and quad8) to reduce the integration subinterval automatically until a given precision is attained. However, if we require the integration of a function whose analytic expression is unknown, MATLAB function **trapz** can be used to perform the numerical integration. The description of the function, **trapz** follows.

The MATLAB **trapz** is used to obtain the numerical integration of a function (with or without an analytic expression) by use of the trapezoidal rules. If a function $f(x)$ has known values at $x_1, x_2, \ldots$ xn given as $f(x_1), f(x_2), \ldots f(x_n)$ respectively, the trapezoidal rule approximates the area under the function, i.e.:

$$A \cong (x_2 - x_1)\left[\frac{f(x_1) + f(x_2)}{2}\right] + (x_3 - x_2)\left[\frac{f(x_2) + (x_3)}{2}\right] + \ldots . \tag{5.13}$$

For constant spacing where

$x_2 - x_1 = x_3 - x_2 = \ldots = h$, the above equation reduces to:

$$A = h\left[\frac{1}{2}f(x_1) + f(x_2) + f(x_3) + \ldots f(x_{n-1}) + \frac{1}{2}f(x_n)\right]. \tag{5.14}$$

The error in the trapezoidal method of numerical integration reduces as the spacing, h, decreases.

The general form of **trapz** function is:

$$S2 = \textbf{trapz(x, y)} \tag{5.15}$$

where

trapz(x, y) computes the integral of y with respect to x. X and y must be vectors of the same length.

Another form of trapz function is:

$$S2 = \textbf{trapz(Y)} \tag{5.16}$$

where

trapz(Y) computes the trapezoidal integral of Y assuming unit spacing data points. If the spacing is different from 1, assuming it is **h**, then trapz(Y) should be multiplied by **h** to obtain the numerical integration. that is:

$$S1 = (h)(S2) = (h).\text{trapz}(Y). \tag{5.17}$$

The following example shows the use of the trapz function.

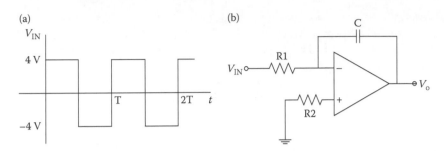

FIGURE 5.4
(a) Square wave input. (b) Op amp integrator.

Example 5.6: Integrator Circuit with a Square Wave Input

A square wave, shown in Figure 5.4a, is applied at the input of an integrator, shown in Figure 5.4b. R1 = R2 = 10 KΩ, C = 1 µF, and the period of the square wave is 2 ms. If the capacitor has zero initial voltage, (a) plot the output waveform, and (b) calculate the root-mean-squared value of the output voltage.

Solution

For the op amp integrator, the output voltage V_0 is given as:

$$V_0(t) = -\frac{1}{RC}\int_0^t V_{IN}(\tau)d\tau. \tag{5.18}$$

Given the input voltage, the trapz function will be used to perform the numerical integration. The rms value of the output waveform is given as:

$$V_{0,rms} = \sqrt{\frac{1}{T_0}\int_0^{T_0} V_0^2 dt}. \tag{5.19}$$

MATLAB® Script

```
% This program calculates the output voltage of
% an integrator
% In addition, we can calculate the rms voltage of the
% output voltage
%
R = 10e3;    C = 1e-6; %values of R and C
T = 2e-3;    %period of square wave
a = 0;   %Lower limit of integration
b = T;   %Upper limit of integration
n = 0:0.005:1; %Number of total data points
```

```
% Obtain output voltage
m = length(n);
% Generate time
for i = 1:m
     t(i) = T*i/m;
   if t(i) < 1e-3
     VX(i) = 4.0;
   else
     VX(i) = -4.0;
   end
     vo_int(i) = trapz(t(1:i), VX(1:i));
     vo(i) = -vo_int(i)/(R*C);        % output voltage
     vo_sq(i) = vo(i)^2; % squared output voltage
end
%
plot(t(1:200), vo(1:200)),   % plot of vo
xlabel('Time in Sec')
Title('Output Voltage, V')
% Determine rms value of output
s = trapz(t(1:m), vo_sq(1:m));   % numerical integration
vo_rms = sqrt(s/b);   % rms value of output
%print out the result
fprintf('rms value of output is %7.3e\n', vo_rms)
The result obtained from MATLAB is;
rms value of output is 2.275e-001.
```

The plot of the output voltage is shown in Figure 5.5.

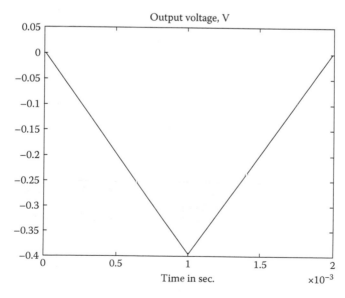

FIGURE 5.5
Output voltage of an op amp integrator.

5.6 Curve Fitting (polyfit, polyval)

The MATLAB® **polyfit** function is used to compute the best fit of a set of data points to a polynomial with a specified degree. The general form of the function is:

$$poly\_xy = polyfit(x, y, n) \qquad (5.20)$$

where
  x and **y** are the data points;
  n is the nth degree polynomial that will fit the vectors x and y; and
  poly_xy is a polynomial that fits the data in vector y to x in the least square sense. **poly_xy** returns $(n + 1)$ coefficients in descending powers of x.

Thus, if the polynomial fit to vectors x and y is given as:

$$Poly\_xy(x) = a_1 x^n + a_2 x^{n-1} + \dots a_m, \qquad (5.21)$$

the degree of the polynomial is **n** and the number of coefficients is **m = n + 1**. The coefficients $(a_1, a_2, \dots a_m)$ are returned by the MATLAB **polyfit** function. An application of the **polyfit** function is illustrated by the following example.

Example 5.7: Zener Diode Parameters from Data

A zener diode at breakdown has the following corresponding voltage and current shown in Table 5.4.
  Plot the graph of current versus voltage. Determine the dynamic resistance.

TABLE 5.4

Voltage and Current of a Zener Diode

Diode Voltage, V v_D	Current, A i_D
−4.686	−1.187E−02
−4.694	−1.582E−02
−4.704	−2.376E−02
−4.708	−2.773E−02
−4.712	−3.170E−02
−4.715	−3.568E−02

Solution

The dynamic resistance of the zener diode is:

$$r_D = \frac{\Delta v_D}{\Delta i_D} \tag{5.22}$$

The plot of v_D versus i_D is almost a straight line with the equation:

$$i_D = m^*v_D + I_0 \tag{5.23}$$

The plot of v_D versus i_D will have the slope given by $1/r_D$. MATLAB® is used to plot the best fit linear model and to calculate resistance of the zener diode.

MATLAB Script

```
%
% Diode parameters
vd = [-4.686 -4.694 -4.699 -4.704 ...
   -4.708 -4.712 -4.715];
id = [-1.187e-002 -1.582e-002 -1.978e-002 ...
   -2.376e-002 -2.773e-002 -3.170e-002 ...
   -3.568e-002];
%
% coefficient
pfit = polyfit (vd, id, 1);
% Linear equation is y = m*x + b
b = pfit(2);
m = pfit(1);
ifit = m*vd + b;
% Calculate Is and n
rd = 1/m
% Plot v versus ln(i) and best fit linear model
plot (vd, ifit, 'b', vd, id, 'ob')
xlabel ('Voltage, V')
ylabel('Diode Current')
title('Best Fit Linear Model')
```

The results obtained from MATLAB are:

rd =
    1.2135

The resistance of the zener diode is 1.2135.

Figure 5.6 shows the best-fit linear model used to determine the resistance of the diode.

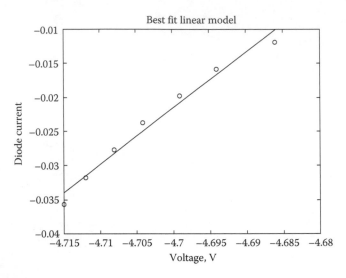

FIGURE 5.6
Voltage versus current of a zener diode.

5.7 Polynomial Functions (roots, poly, polyval, and fzero)

5.7.1 Roots of Polynomials (roots, poly, polyval)

If $f(x)$ is a polynomial of the form:

$$f(x) = C_0 x^n + C_1 x^{n-1} + \dots + C_{n-1} x + C_n \qquad (5.24)$$

$f(x)$ is of degree n and has exactly n roots. The n roots may have multiple **roots** or have complex roots. If the coefficients of the polynomial (C_0, C_1, C_2, ..., C_m) of the polynomial are real, then the complex roots will occur as complex conjugate pairs.

The MATLAB® function for determining the roots of a polynomial is the **roots** function. The general format for using **roots** is:

$$\textbf{roots(c)} \qquad (5.25)$$

where
    c is a vector that contains coefficients of the polynomial (coefficients ordered in descending powers of x).

For example, for a polynomial:

$$g(x) = x^4 + 3x^3 + 2x + 4 = 0 \qquad (5.26)$$

the coefficient c = [1   3   0   2   4]
The roots are obtained by the statement:

```
b = roots(c)
```

The resulting roots are:

```
b =
    -3.0739
    0.5370 + 1.0064i
    0.5370 -  1.0064i
    -1.0000
```

If the roots of a polynomial are known and we want to determine the coefficients of the polynomial that corresponds to the roots, we can use the **poly** function. The general form for using **poly** function is:

$$\textbf{poly(r)} \tag{5.27}$$

where
  r is a vector that contains the results of a polynomial; and
  poly(r) returns the coefficients of the polynomial whose roots are contained in the vector r.

In the previous example, the roots of the polynomial $x^4 + 3x^3 + 2x + 4 = 0$ were:

```
b =
    -3.0739
    0.5370 + 1.0064i
    0.5370 -  1.0064i
    -1.0000
```

To confirm that the roots will give us the corresponding polynomial, we use the statement:

```
g_coeff = poly(b')
```

and we obtain

```
g_coeff =
        1.0000   3.0000   0.0000   2.0000   4.0000
```

It should be noted that the g_coeff are the same as the coefficients of the polynomial g(x) of Equation 5.26.

The MATLAB function **polyval** is used for polynomial evaluation. The general form of polynomial is:

$$\text{polyval}(p, x) \qquad\qquad (5.28)$$

where
    p is a vector whose elements are the coefficients of a polynomial in descending powers; and
    polyval(p, t) is the value of the polynomial evaluated at x.

For example, to evaluate the polynomial:

$$h(x) = 3x^4 + 4x^3 + 5x^2 + 2x + 1 \qquad\qquad (5.29)$$

at x = 3, we use the statements

```
p = [3  4  5  2  1]
polyval (p, 3)
```

Then we get:

```
ans =
    403
```

5.7.2 Zero of a function (fzero) and nonzero of a function (find)

The MATLAB® function **fzero** is used to obtain the zero of a function of one variable. The general form of the **fzero** function is:

$$\text{fzero('function', x1)}$$

$$\text{fzero('function', x1, tol)}$$

where
    fzero('funct', x1) finds the zero of the function funct(x) that is near the point x1; and
    fzero('funct', x1, tol) returns zero of the function funct(x) accurate to the relative error of *tol*.

The **find** function determines the indices of the nonzero elements of a vector or matrix. The statement:

$$C = \text{find}(f)$$

will return the indices of the vector f that are nonzero. For example, to obtain the points where a change of sign occurs, the statement:

```
D = find(product < 0)
```

will show the indices of the locations in product that are negative.

5.7.3 Frequency Response of a Transfer Function (freqs)

MATLAB® function **freqs** is used to obtain the frequency response function H(s). The general form of the function is:

$$hs = \text{freqs(num, den, range)} \tag{5.30}$$

$$H(s) = \frac{y(s)}{x(s)} = \frac{b_m s^m + b_{m-1} s^{m-1} + \ldots + b_1 s + b_0}{a_n s^n + a_{n-1} s^{n-1} + \ldots + a_1 s + a_0}; \tag{5.31}$$

where
num = $[b_m \ b_{m-1} \ b_1 \ b_0]$, coefficients of numerator polynomial;
den = $[a_n \ a_{n-1} \ldots a_1 \ a_0]$, coefficients of denominator polynomials;
range is the range of frequencies; and
hs is frequency response (in complex number form).

freqs is an m-file in the MATLAB Signal Processing Toolbox. It is also available in the student edition of MATLAB. The polynomial at each frequency point is evaluated. It then divides the numerator response by the denominator response. The **freqs** algorithm is:

```
s = sqrt(-1)*w;
h = polyval(b,s)./ polyval(a,s);
```

The following example explores the use of the **freqs** function.

Example 5.8: Frequency Response from Transfer Function

Determine the magnitude response of a system whose transfer function is given as:

$$H(s) = \frac{4s}{s^2 + 64s + 16}. \tag{5.32}$$

Solution
We can use the following MATLAB® script to obtain the magnitude response.

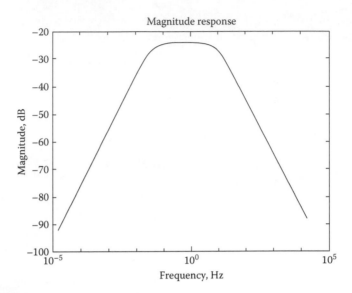

FIGURE 5.7
The magnitude response from a transfer function.

MATLAB Script

```
%
% Magnitude response of a transfer function
num = [4 0]; % coefficients of numerator polynomial
den = [1 64 16]; % coefficients of denominator polynomial
w = logspace (-4, 5); % range of frequencies
hs = freqs(num, den, w);
f = w/(2*pi); % frequency in Hz.
hs_mag = 20*log10(abs(hs)); % Magnitude in decibels
% Plot the magnitude response
semilogx(f, hs_mag)
title ('Magnitude Response')
xlabel('Frequency, Hz')
ylabel('Magnitude, dB')
```

The magnitude response is shown in Figure 5.7.

The following example uses the MATLAB functions **freqs** and **find** to obtain unity gain frequency of an amplifier.

Example 5.9: Unity Gain Crossover Frequency

A block diagram of an amplifier is shown in Figure 5.8. The unity gain crossover frequency is the frequency wherein the magnitude of a transfer function is unity. If the transfer function of the amplifier is given as:

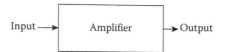

FIGURE 5.8
Block diagram of an amplifier.

$$H(s) = \frac{2.62 \times 10^{18}}{(s + 400\pi)(s + 8\pi \times 10^5)(s + 1.6\pi \times 10^6)} \tag{5.33}$$

Determine the crossover frequency.

Solution

MATLAB® Script

```
% Gain crossover frequency
% Transfer function parameters
% poles are
p1 = 400*pi; p2 = 8e5*pi; p3 = 1.6e6*pi;
% determine the coefficients for numerator
% and denominator polynomial
a2 = p1 + p2 + p3;
a1 = p1*p2 + p1*p3 + p2*p3;
a0 = p1*p2*p3;
den = [1 a2 a1 a0];    % coefficients of denominator
polynomial
num = [2.62e18];       % coefficients. of numerator
polynomial
w = logspace(-1, 7, 5000);    % range of frequencies
hs = freqs(num, den, w);
hs_mag = 20*log10(abs(hs));   % magnitude characteristics
%
f = w/(2*pi);
plot(f,hs_mag)
xlabel('Frequency, Hz')
ylabel('Gain, dB')
title('Frequency Response of an Amplifier')
% gain crossover calculation, unity gain = 0 db gain
lenw = length(w);
lenw1 = lenw - 1;
for i = 1:lenw1
  prod(i) = hs_mag(i)*hs_mag(i + 1);
end
fcrit = f(find(prod < 0));
f_cross = fcrit;
fprintf('The crossover frequency is % 9.4e\n', f_cross)
The result obtained from MATLAB is;
The crossover frequency is 3.2861e + 004 Hz.
```

The plot of the magnitude response is shown in Figure 5.9.

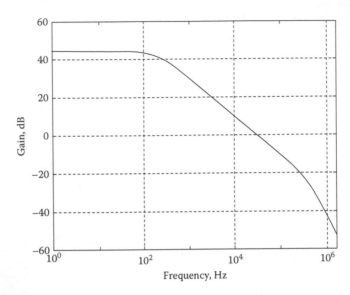

FIGURE 5.9
Magnitude response of an amplifier.

5.8 Save, Load, and Textread Functions

In Section 4.9, some input/output commands of MATLAB® were discussed. These were the break, disp, echo, format, fprintf, input, and pause functions. In this section, additional input/output commands will be discussed. These are save, load, and textread functions.

5.8.1 Save and Load Commands

The **save** command saves data in MATLAB® workspace to disk. The **save** command can store data either in memory-efficient binary format, called a MAT-file, or ASCII file. The general form of the **save** command is:

save filename [List of variables] [options]

where
   save (without filename, list of variables, and options), saves all the data
       in the current workspace to a file named **matlab.mat** in the current
       directory;
   If a filename is included in the command line, the data will be saved in file
       "filename.mat"; and
   If a list of variables is included, only those variables will be saved.

The options for the save command are shown in Table 5.5.

TABLE 5.5

Save Command Options

Option	Description
-mat	Save data in MAT-file format (default)
-ascii	Save data using 8-digit ASCII format
-ascii -double	Save data using 16-digit ASCII format
-ascii -double -tab	Saves data using 16-digit ASCII format with Tabs
-append	Save data to an existing MAT-file
-v4	Save data in a format that MATLAB® Version 4 can open and read

TABLE 5.6

Load Command Option

Option	Description
-mat	Load data from MAT-file (default in file extension is mat)
-ascii	Load data from space – separated file

MAT-files are preferable for data that are generated and are going to be used by MATLAB. MAT-files are platform-independent. The files can be written and read by any computer that supports MATLAB. In addition, MAT-files preserve all the information about each variable in the workspace including its name, size, storage space in bytes, and class (structure array, double array, cell array, cell array, or character array). Furthermore, MAT-files have every variable stored in full precision.

The **ASCII files** are preferable if the data is to be exported or imported to programs other than MATLAB. It is recommended that if you save workspace content in ASCII format, *save only one variable at a time*. If more than one variable is saved, MATLAB will create the ASCII data that might be difficult to interpret when loaded back in a MATLAB program.

The **load** command will load data from MAT-file or ASCII file into the current workspace. The general format of the **load** command is:

load filename [options]

where

load (by itself without filename and options) will load all the data in file matlab.mat into the current workspace; and

load filename – will load data from the specified filename.

The options for the load command are shown in Table 5.6.

It is *strongly recommended* that an ASCII data file, which will be used with MATLAB program, should contain only numeric information and each row of the file should have the same number of data values. It is also recommended that an ASCII filename include the extension **.dat** so that it may be easier to distinguish m-files and MAT-files.

TABLE 5.7

Data Stored in File rc_1.dat

Time, s	Voltage, V
0.0	0.0
0.5	3.94
1.0	6.32
1.5	7.77
2.0	8.65
2.5	9.18
3.0	9.50
3.5	9.69
4.0	9.82
4.5	9.89
5.0	9.93

Suppose that a data file stored on disk under a filename **rc_1.dat** contains the data shown in Table 5.7.

The following command:

```
load    rc_1.dat -ascii
```

will load the data into MATLAB and the data will be stored in the matrix rc_1, which has two columns of data. If one gives the command:

```
rc_1
```

then, one will get

```
rc_1 =
     0          0
     0.5000    3.9400
     1.0000    6.3200
     1.5000    7.7700
     2.0000    8.6500
     2.5000    9.1800
     3.0000    9.5000
     3.5000    9.6900
     4.0000    9.8200
     4.5000    9.8900
     5.0000    9.9300.
```

5.8.2 The Textread Function

The **textread** command can be used to read ASCII files that are formatted into columns of data, where values in each column might be a different type. The general form of the **textread** command is:

$$[a, b, c, \ldots] = \text{textread(filename, format, n)}$$

where

filename is the name of file to open. The filename should be in quotes, i.e. 'filename';

format is a string containing a description of the type of data in each column. The format descriptors are similar to those of fprintf, discussed in Section 3.9. The format list should be in quotes.

Supported functions include:

%d – read a signed integer value

%u – read an integer value

%f – read a floating point value

%s – read a whitespace separated string

%q – read a (possibly double quoted) string

%c – read characters (including white space) (output is char array);

n is the number of lines to read. If *n* is missing, the command reads to the end of the file; and

a,b,c… are the output arguments. The number of output arguments must match the number of columns that are being read.

The **textread** is much more than the **load** command. The **load** command assumes that all the data in the file being loaded is of a single type. The **load** command does not support different data types in different columns. In addition, the **load** command stores all the data in a single array. But, the **textread** command allows each column of data to go into a separate variable.

For example, suppose the file rc_2.dat contains the data shown in Table 5.8. The first column is time and the second column the voltages across a capacitor, we can use **textread** function to read the data.

```
[time,volt_cap] = textread('rc_2.dat', '%f %f')
time
volt_cap
```

TABLE 5.8

Data Stored in File rc_2.dat

Time, s	Voltage
0.0	50.0
1.0	30.3
2.0	18.4
3.0	11.2
4.0	6.77
5.0	4.10
6.0	2.49
7.0	1.51
8.0	0.916

If the above statements are executed, the results are:

```
time =
         0
         1
         2
         3
         4
         5
         6
         7
         8

volt_cap =
        50.0000
        30.3000
        18.4000
        11.2000
         6.7700
         4.1000
         2.4900
         1.5100
         0.9160
```

The following example will illustrate the use of the **load** function.

Example 5.10: Statistical Analysis of Data Stored in File

The data shown in Table 5.9 represent two voltages obtained from Monte Carlo analysis of a circuit. V1 and V2 are voltages at two nodes of the circuit. The data is stored in file fproc.dat. (a) Read data from file, plot V1 as a function of simulation run. (b) Find the mean and standard deviation of V1 and V2.

Solution

Since the **textread** function is not applicable to data in exponential notation, the **load** command will be used to read in the data on file.

MATLAB® Script

```
% data is stored in fproc.dat
% read data using load command
%
load fproc.dat -ascii
k = fproc(:,1);
v1 = fproc(:,2);
v2 = fproc(:,3);
n = length(k);
% calculate the mean and standard deviation
mean_v1 = mean(v1);    % mean of V1
```

```
std_v1 = std(v1);      % standard deviation of v1
mean_v2 = mean(v2);    % mean of V2
std_v2 = std(v2);      % standard deviation of V2
plot(k, v1); % plot of simulation run and V1
xlabel ('Simulation Run')
ylabel('Voltage V1')
title('Plot of Voltage V1')
% Print out results
fprintf('Mean value of V1 is%9.4e volts\n',mean_v1)
fprintf('Standard deviation of V1 is %9.4e volts\n', std_v1)
fprintf('Mean value of V2 is %9.4e volts\n', mean_v2)
fprintf('Standard deviation of V2 is %9.4evolts\n', std_v2)
```

The results are:

Mean value of V1 is 3.4784e + 000 volts;
Standard deviation of V1 is 2.3638e-001 volts;
Mean value of V2 is 8.2100e-001 volts; and
Standard deviation of V2 is 5.3792e-002 volts

Figure 5.10 shows the voltage V1 as a function of the simulation run.

TABLE 5.9

Voltages Obtained from Monte Carlo Analysis

Simulation Run	Voltage, V1	Voltage, V2
1	3.393E+00	8.262E–01
2	3.931E+00	8.483E–01
3	3.761E+00	7.991E–01
4	3.515E+00	8.877E–01
5	3.716E+00	8.922E–01
6	3.243E+00	8.267E–01
7	3.684E+00	7.838E–01
8	3.314E+00	7.687E–01
9	3.778E+00	7.661E–01
10	3.335E+00	9.185E–01
11	3.332E+00	7.991E–01
12	2.993E+00	8.460E–01
13	3.505E+00	7.274E–01
14	3.380E+00	7.873E–01
15	3.584E+00	9.163E–01
16	3.697E+00	7.829E–01
17	3.373E+00	8.119E–01
18	3.106E+00	8.082E–01
19	3.453E+00	7.590E–01
20	3.474E+00	8.647E–01

FIGURE 5.10
Voltage V1 as a function of simulation run.

5.9 Interfacing SPICE to MATLAB®

As mentioned in Chapter 1, SPICE is the defacto standard for circuit simulation. It can perform DC, AC, transient, Fourier, and Monte Carlo analysis. In addition, SPICE has device models incorporated in its package. Furthermore, there are extensive libraries of device models available that a SPICE program user can use for simulation and design. As discussed in Section 3.6, PSPICE has analog behavioral model facilities that allow modeling of analog circuit functions by using mathematical equations, tables, and transfer functions. The above features of PSPICE are unmatched by other scientific packages, such as MATLAB®, MATHCAD, and MATHEMATICA. On the other hand, MATLAB is primarily a tool for matrix computations. It has numerous functions for data processing and analysis. In addition MATLAB has a rich set of plotting capabilities, which is integrated into the MATLAB package. Furthermore, since MATLAB is also a programming environment, a user can extend the MATLAB functional capabilities by writing new modules (m_files).

The book uses the strong features of PSPICE and the powerful functions of MATLAB for Electronic circuit analysis. PSPICE can be used to perform DC, AC, transient, Fourier, temperature, and Monte Carlo analysis of electronic circuits with device models and subsystem sub-circuits. Then, MATLAB can be used to perform calculations of device parameters, curve fitting, numerical integration, numerical differentiation, statistical analysis, and two-dimensional and three-dimensional plots.

PSPICE has the postprocessor package, **PROBE** that can be used for plotting PSPICE results. PROBE also has built-in functions that can be used to do simple signal processing. The valid functions for PROBE expressions are shown in Table 1.4. Compare Table 1.4 of the PROBE expression and Tables 5.1 and 5.2 of the MATLAB mathematical and data analysis functions. PSPICE PROBE mathematical expressions do not have the MATLAB functions shown in Table 5.10.

It can be seen from Table 5.10 that MATLAB has extensive functions for data analysis, unavailable in PSPICE.

Both PSPICE and MATLAB have functions for performing numerical integration and differentiation. These functions are shown in Table 5.11.

TABLE 5.10

MATLAB Functions Unavailable in PSPICE

MATLAB Function	Description
corrcoeff	Obtains correlation coefficients
cov(x)	Determines covariance matrix
cross(x,y)	Finds cross product of vectors x and y
cumprod(x)	Obtains the cumulative product of columns
cumsum(x)	Obtains the cumulative sum of columns or cumulative sum of elements in a column
hist(x)	Draws the histogram or the bar chart of x using 10 bins
median(x)	Finds the median value of the elements in the vector x
std(x)	Calculates and returns the standard deviation of x
rand(x)	Produces an n-by-n matrix containing normally distributed (Gaussian) random numbers with a mean of zero and variance of 1.
sort(x)	Sort the rows of a matrix a in ascending order
sum(x)	Calculates and returns the sum of the elements in x
fzero(x)	Finds zero of a function
find(x)	Determines the indices of the nonzero elements of x
polyfit	Determines the polynomial curve fit
fix(x)	Round x to the nearest integer toward zero
floor(x)	Rounds x to the nearest integer toward minus infinity
round(x)	Round toward nearest integer

TABLE 5.11

Functions for Numerical Integration and Differentiation in PSPICE and MATLAB

Mathematical Operation	PSPICE PROBE Function	MATLAB Function
Numerical integration	s(x)	quad('funct', a, b, tol, trace) quad8('funct', a, b, tol, trace) S2 = trapz(x, y)
Numerical differentiation	d(x)	$f'(x) \cong \dfrac{\text{diff}(f)}{\text{diff}(x)}$

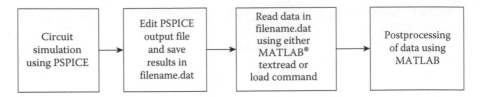

FIGURE 5.11
Flowchart of circuit simulation using PSPICE and postprocessing of PSPICE.

However, MATLAB has several functions for numerical integration. Some of the MATLAB functions allow the user to specify the tolerance limit for stopping the numerical integration (tol). This facility is unavailable in PSPICE.

To exploit the best features of PSPICE and MATLAB, circuit simulations will be done using PSPICE. The PSPICE results, which are written into a file named, **filename.out,** will be edited using a text editor or a word processor and the data will be saved as **filename.dat**. The data will be read using either MATLAB **textread** or **load** commands. Further processing on the data will be done using MATLAB. The methodology is shown in Figure 5.11.

The following chapters use the methodology described in this section to analyze electronic circuits.

Problems

5.1   A triangular wave, symmetric about $t = 0$ has a Fourier series expansion:

$$x(t) = \frac{8A}{\pi^2}\left(\cos \omega t + \frac{1}{9}\cos 3\omega t + \frac{1}{25}\cos 5\omega t + \frac{1}{49}\cos 7\omega t + \ldots\right)$$

where A is peak value and ω is the angular frequency. If $A = 2$, $\omega = 400\pi$ rad/s, write a MATLAB® program to synthesize the triangular wave using the first 4 terms.

5.2   A square wave with a peak to peak value of 4 V and an average value of zero volts can be expressed as:

$$x(t) = \frac{8}{\pi}\left(\cos \omega_0 t - \frac{1}{3}\cos 3\omega_0 t + \frac{1}{5}\cos 5\omega_0 t - \frac{1}{7}\cos 7\omega_0 t + \ldots\right)$$

where

$$\omega = 2\pi f_0 \text{ and } f_0 = 1000 Hz.$$

Write a MATLAB program to synthesize the square wave using the first 4 terms.

5.3   The gains β of fifteen 2N2907 transistors, measured in the laboratory are 170, 200, 160, 165, 175, 155, 210, 190, 180, 165, 195, 200, 195, 205, and 190.

a. What is the mean value of β?

b. What are the minimum and maximum values of β?

c. What is the standard deviation of β?

5.4 The zener voltage regulator circuit, shown in Figure P5.4, has RS = 250 Ω and RL = 350 Ω. The diode D1 is D1N750. The corresponding input and output voltages are shown in Table P5.4.

a. Find the minimum, maximum, mean, and standard deviation of the input and output voltages.

b. What is the correlation coefficient between input and output voltages?

c. What is the least mean square fit (polynomial fit)?

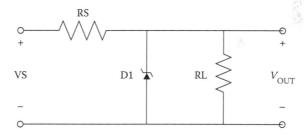

FIGURE P5.4

Zener diode shunt voltage regulator circuit.

TABLE P5.4

Input and Output Voltages of Zener Voltage Regulator

Input Voltage VS, V	Output Voltage V_{OUT}, V
8.0	4.601
9.0	4.658
10.0	4.676
11.0	4.686
12.0	4.694
13.0	4.699
14.0	4.704
15.0	4.708
16.0	4.712
17.0	4.715
18.0	4.717

5.5 The transfer function of a bandpass filter is:

$$H(s) = \frac{s\dfrac{R}{L}}{s^2 + s\dfrac{R}{L} + \dfrac{1}{LC}}.$$

If L = 5 mH, C = 20μF and R = 15 KΩ:

a. Use MATLAB to plot the magnitude response; and
b. Find the frequency at which the maximum value of the magnitude response occurs.

5.6   For the differentiator circuit shown in Figure P5.6a, has R = 10 K and C = 10 F. The input voltage is sawtooth waveform shown in Figure P5.6b. Use MATLAB to plot the output waveform. Calculate the mean value and rms value of the output voltage, $v_0(t)$.

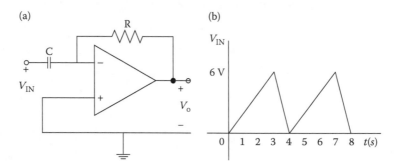

(a)

(b)

FIGURE P5.6

(a) Integrator circuit. (b) Sawtooth waveform.

5.7   A static random access memory (SRAM) operating in a radiation environment can experience a change in logic state. If currents at various instants of time at a node of a SRAM are shown in Table P5.7, calculate the charge deposited on the node.

TABLE P5.7

Current at SRAM Node

Time, s	Current, A
0.000E+00	0.000E+00
2.000E–08	0.000E+00
4.000E–08	4.000E–12
6.000E–08	3.672E–02
8.000E–08	3.970E–02
1.000E–07	3.998E–02
1.200E–07	4.000E–02
1.400E–07	4.000E–02
1.600E–07	3.281E–02
1.800E–07	2.200E–02
2.000E–07	1.475E–02
2.200E–07	9.887E–03
2.400E–07	6.628E–03

TABLE P5.7

Current at SRAM Node (Continued)

Time, s	Current, A
2.600E–07	4.442E–03
2.800E–07	2.978E–03
3.000E–07	1.996E–03
3.200E–07	1.338E–03
3.400E–07	8.969E–04
3.600E–07	6.013E–04
3.800E–07	4.030E–04
4.000E–07	2.695E–04

5.8   In a platinum resistance thermometer with the American Alloy, the resistance, R, is related to the temperature, T, by the expression:

$$R = a + bT + cT^2$$

where R is the resistance in Ohms, T is temperature in °C. Using the data in Table P5.8, determine the coefficients a, b, and c.

TABLE P5.8

Temperature Versus Resistance of a Thermometer

T, °C	R, Ω
−100	60.30
−80	68.34
−60	76.32
−40	84.26
−20	92.16
0	100
20	107.79
40	115.54
60	123.24
80	130.89
100	138.50

5.9   To the first order in 1/T, the relationship between resistance R and temperature T of a thermister is given by:

$$R(T) = R(T_0)\exp\left(\beta\left(\frac{1}{T} - \frac{1}{T_0}\right)\right)$$

where
T is in degrees Kelvin;
T_0 is the reference temperature in degrees Kelvin;

$R(T_0)$ is the resistance at the reference temperature; and β is the temperature coefficient.

If $R(T_0)$ is 10,000 Ω at the reference temperature, use Table P5.9 to obtain the constants β and T_0.

TABLE P5.9

Temperature Versus Resistance of a Thermister

T, °K	R, Ω
240	1.7088E+005
260	0.5565E+005
280	0.2128E+005
300	0.0925E+005
320	0.0446E+005
340	0.0234E+005
360	0.0132E+005
380	0.0079E+005
400	0.0050E+005
420	0.0033E+005
440	0.0023E+005

5.10 An amplifier has the following voltage transfer function:

$$A(s) = \frac{150s}{\left(1 + \dfrac{s}{10^3}\right)\left(1 + \dfrac{s}{5 \times 10^4}\right)}$$

a. Plots the magnitude versus frequency; and
b. Find the gain-crossover frequency, fgc, where the magnitude becomes unity.

5.11 The voltage transfer function of an uncompensated op amp is given as:

$$A(s) = \frac{250}{\left(1 + \dfrac{s}{200\pi}\right)} \frac{100}{\left(1 + \dfrac{s}{80\pi}\right)} \frac{0.8}{\left(1 + \dfrac{s}{2.5\pi \times 10^7}\right)}$$

a. Draw the Bode plot for the magnitude versus frequency; and
b. Find the gain-crossover frequency, fgc, where the magnitude of A(s) becomes unity.

5.12 The voltage between two nodes of circuit for different supply voltages are shown in Table P5.12. Plot V1 versus V2. Determine the line of best fit between V1 and V2.

TABLE P5.12

Voltages at Two Nodes

Voltage V1, V	Voltage V2, V
5.638	5.294
5.875	5.644
6.111	5.835
6.348	6.165
6.584	6.374
6.820	6.684
7.055	6.843
7.290	7.162
7.525	7.460
7.759	7.627
7.990	7.972
8.216	8.170
8.345	8.362

5.13 A rectangular waveform can be expressed as:

$$x(t) = \frac{8}{\pi}\left(\cos\omega_0 t - \frac{1}{3}\cos 3\omega_0 t + \frac{1}{5}\cos 5\omega_0 t - \frac{1}{7}\cos 7\omega_0 t + \dots \right)$$

where

$$\omega = 2\pi f_0 \quad \text{and} \quad f_0 = 5000 Hz.$$

Write a MATLAB program to synthesize the square wave using the first 6 terms.

5.14 The resistances measured from an electronics laboratory had the following values: 121, 119, 117, 122, 118, 124, 121, 116, 121, and 119.
   a. What is the mean value of the resistors?
   b. What is the median of the resistance values?
   c. What percentage of resistors have resistance values greater than $|120 \pm 2\%|$?

5.15 The transfer function of a filter is:

$$H(s) = \frac{s^2 + 19.8 \times 10^4}{s^2 + 648s + 28.7 \times 10^4}.$$

   a. Use MATLAB to plot the magnitude response.
   b. Determine the center frequency.
   c. Find the bandwidth of the filter.

Bibliography

1. Attia, J. O. *Electronics and Circuit Analysis Using MATLAB®*. 2nd ed. Boca Raton, FL: CRC Press, 2004.
2. Biran, A., and M. Breiner. *MATLAB® for Engineers*. White Plains, NY: Addison-Wesley, 1995.
3. Boyd, Robert R. *Tolerance Analysis of Electronic Circuits Using MATLAB®*. Boca Raton, FL: CRC Press, 1999.
4. Chapman, S. J. *MATLAB® Programming for Engineers*. Tampa, FL: Thompson, 2005.
5. Davis, Timothy A., and K. Sigmor. *MATLAB® Primer*. Boca Raton, FL: Chapman & Hall/CRC, 2005.
6. Etter, D. M. *Engineering Problem Solving with MATLAB®*. 2nd ed. Upper Saddle River, NJ: Prentice Hall, 1997.
7. Etter, D. M., D. C. Kuncicky, and D. Hull. *Introduction to MATLAB® 6*. Upper Saddle River, NJ: Prentice Hall, 2002.
8. Gilat, Amos. *MATLAB®, An Introduction With Applications*. 2nd ed. New York: John Wiley & Sons, 2005.
9. Gottling, J. G. *Matrix Analysis of Circuits Using MATLAB®*. Upper Saddle River, NJ: Prentice Hall, 1995.
10. Hahn, Brian D., and Daniel T. Valentine. *Essential MATLAB® for Engineers and Scientists*. 3rd ed. New York and London: Elsevier, 2007.
11. Herniter, Marc E. *Programming in MATLAB®*. Florence, KY: Brooks/Cole Thompson Learning, 2001.
12. Howe, Roger T., and Charles G. Sodini. *Microelectronics, An Integrated Approach*. Upper Saddle River, NJ: Prentice Hall, 1997.
13. Moore, Holly. *MATLAB® for Engineers*. Upper Saddle River, NJ: Pearson Prentice Hall, 2007.
14. *Using MATLAB®, The Language of Technical Computing, Computation, Visualization, Programming, Version 6*. Natick, MA: MathWorks, Inc. 2000.

Part III

6

Diode Circuits

In this chapter, we discuss diode circuits. The chapter begins with diode characteristics, followed by diode rectification. Peak detector and limiter circuits that use diodes are also discussed. Zener diodes and zener voltage regulator circuits are also covered. Most of the examples in the chapter are done using both PSPICE and MATLAB®.

6.1 Diode

In the forward-biased and reversed-biased regions, the current, i_D, and the voltage, v_D, of a semiconductor diode are related by the diode equation:

$$i_D = I_s[e^{(v_D/nV_T)} - 1] \qquad (6.1)$$

where
 I_s is reverse saturation current;
 n is an empirical constant between 1 and 2; and
 V_T is thermal voltage, given by

$$V_T = \frac{kT}{q} \qquad (6.2)$$

and
 k is Boltzman's constant = 1.38×10^{-23} J/°K;
 q is the electronic charge = 1.6×10^{-19} coulombs; and
 T is the absolute temperature in °K.

In the forward-biased region, the voltage across the diode is positive. Assuming that the voltage across the diode is greater than 0.5 V, Equation 6.1 simplifies to:

$$i_D = I_s e^{v_D/nV_T} \qquad (6.3)$$

From Equation 6.3, we get:

$$\ln(i_D) = \frac{v_D}{nV_T} + \ln(I_S) \tag{6.4}$$

For a particular operating point of the diode ($i_D = I_D$, and $v_D = V_D$,), we can obtain the dynamic resistance of the diode, r_d, at a specified operating point as:

$$r_d = \frac{di_D}{dv_D}\bigg|_{v_D=V_D} = \frac{I_S e^{(v_D/nV_T)}}{nV_T} \tag{6.5}$$

Equation 6.4 can be used to obtain the diode constants n and I_s, given a data that consists of the corresponding values of voltage and current. From Equation 6.4, a curve of v_D versus $\ln(i_D)$ will have a slope given by $1/nV_T$ and a y-intercept of $\ln(I_S)$. The following example illustrates how to find n and I_S from experimental data. The example uses the MATLAB® function **polyfit**, which was discussed in Chapter 4.

Example 6.1: Determination of Diode Parameters from Data

A forward-biased, semiconductor diode has the following corresponding voltage and current shown in Table 6.1. Determine the reverse saturation current I_S and the diode constant n. Plot the line of best fit.

Solution

MATLAB® Script

```
% Diode parameters
vt = 25.67e-3;
vd = [0.2 0.3 0.4 0.5 0.6 0.7];
id = [6.37e-9 7.75e-8 6.79e-7 3.97e-6 5.59e-5 3.63e-4];
%
lnid = log(id);     % Natural log of current
% Determine coefficients
pfit = polyfit (vd, lnid, 1);% curve fitting
% Linear equation is y = mx + b
b = pfit(2);
m = pfit(1);
ifit = m*vd + b;
% Calculate Is and n
Is = exp(b)
n = 1/(m*vt)
% Plot v versys ln(i) and best fit linear model
plot(vd, ifit, 'b', vd, lnid, 'ob')
xlabel ('Voltage, V')
ylabel ('ln(i)')
title ('Best Fit Linear Model')
```

TABLE 6.1

Current Versus Voltage of a
Forward-Biased Diode

Forward-Biased Voltage v_D, V	Forward-Biased Current i_D, A
0.2	6.37e−9
0.3	7.75e−8
0.4	6.79e−7
0.5	3.97e−6
0.6	5.59e−5
0.7	3.63e−4

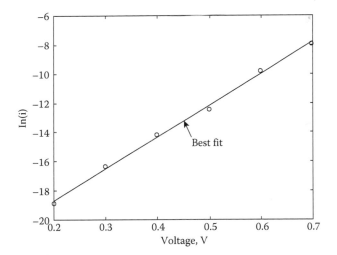

FIGURE 6.1
Best fit linear model.

The results obtained from MATLAB are:

$$Is =$$

$$9.5559e{-}011$$

$$n =$$

$$1.7879$$

Figure 6.1 shows the best fit linear model used to determine the reverse saturation current I_s and the diode parameter n.

In the diode Equation 6.1, the thermal voltage V_T and the reverse saturation current I_s are temperature dependent. The thermal voltage is directly proportional to temperature. This is shown in Equation 6.2. The reverse saturation current increases approximately 7.2%/°C for both silicon and germanium diodes. The expression for the reverse saturation current as a function of temperature is:

$$I_S(T_2) = I_S(T_1)e^{[k_s(T_2-T_1)]}$$ (6.6)

where
$K_S = 0.072/°C$; and
T_1 and T_2 are two different temperatures.

The following example shows the effect of the temperature on the output voltage of a diode circuit.

Example 6.2: Temperature Effects on a Diode

For the circuit shown in Figure 6.2, VS = 5 V, R1 = 50 KΩ, R2 = 20 KΩ, and R3 = 50 KΩ. If the diode D1 is 1N4009, plot temperature versus output voltage. Find the equation of best fit between the voltage and the temperature.

Solution

PSPICE is used to obtain the voltage at various temperatures. The SPICE command .TEMP is used to obtain the temperature effects. MATLAB® is used to plot the relationship between temperature and diode voltage.

PSPICE Program

```
DIODE CIRCUIT
VS     1     0     DC     5V
R1     1     2     50E3
R2     2     0     20E3
R3     2     3     50E3
D1     3     0     1N4009
.MODEL 1N4009  D(IS = 0.1P RS = 4 CJO = 2P TT = 3N BV = 60
IBV = 0.1P)
.STEP          TEMP  0      100    10
.DC    VS            5      5      1
.PRINT  DC   V(3)
.END
```

The PSPICE results showing the temperature versus voltage across the diode can be found in file ex5_2ps.dat. The results are shown in Table 6.2.
The MATLAB program for plotting the voltage versus temperature.

MATLAB Script

```
% Processing of PSPICE data using MATLAB
% Read data using textread command
%
[temp, vdiode] = textread ('ex5_2ps.dat', '%d %f');
vfit = polyfit(temp,vdiode, 1);
% Linear equation is y = mx + b
```

```
b = vfit(2)
m = vfit(1)
vfit = m*temp + b;
plot(temp, vfit, 'b', temp, vdiode, 'ob');% plot
temperature vs. diode voltage
xlabel ('Temperature in °C')
ylabel ('Voltage, V')
title ('Temperature versus Diode Voltage')
```

From the MATLAB program,

$$b =$$

0.5480

$$m =$$

−0.0023

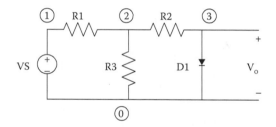

FIGURE 6.2
Diode circuit.

TABLE 6.2

Diode Voltage versus Temperature

Temperature, °C	Diode Voltage, V
0	0.5476
10	0.5250
20	0.5023
30	0.4796
40	0.4568
50	0.4339
60	0.4110
70	0.3881
80	0.3651
90	0.3421
100	0.3190

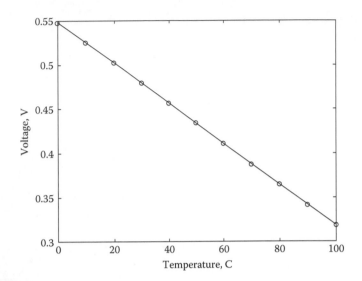

FIGURE 6.3
Temperature versus diode voltage.

The equation of best fit between temperature and voltage is:

$$v_O(T) = -0.0023T + 0.548 \text{ V}.$$

The plot is shown in Figure 6.3.

6.2 Rectification

A half-wave rectifier circuit is shown in Figure 6.4. It consists of an alternating current (AC) source, a diode, and a resistor. Assuming that the diode is ideal, the diode conducts when the source is positive, making:

$$v_0 = v_S \quad \text{when} \quad v_S > 0. \tag{6.7}$$

When the source voltage is negative, the diode is cut-off, and the output voltage is:

$$v_0 = 0 \quad \text{when} \quad v_S < 0. \tag{6.8}$$

A battery charging circuit is a slight modification of the half-wave rectifier circuit. The battery charging circuit, explored in the following example,

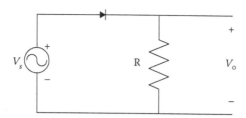

FIGURE 6.4
Half-wave rectifier circuit.

consists of an alternating current source connected to a battery through a resistor and a diode.

Example 6.3: Battery Charging Circuit

In the battery charging circuit shown in Figure 6.5, VB = 12 V and R = 50 Ω. The source voltage is $v_s(t) = 16\sin(120\pi t)$ V. If the diode is D1N4448, (a) plot the current flowing through the diode, (b) determine peak current and average current flowing into the battery, and (c) what is the total charge supplied to the battery?

Solution

PSPICE is used to obtain the output current as a function of time. Three cycles of the input signal is used in the simulations. The resulting data are analyzed using MATLAB®.

PSPICE Program

```
* BATTERY CHARGING CIRCUIT
VS      1      0      SIN(0  16  60)
R       1      2      50
D       2      3      D1N4448
VB      3      0      DC    12V
.MODEL D1N4448 D(IS=0.1P RS=2 CJO=2P TT=12N BV=100 IBV=0.1P)
.TRAN 0.5MS     50MS     0     0.5MS
.PRINT          TRAN    I(R)
.PROBE
.END
```

The PSPICE results showing the current flowing through the diode can be found in file ex6_3ps.dat. The partial results are in shown in Table 6.3.

The MATLAB program for the analysis of the PSPICE results follows:

MATLAB Script

```
% Battery charging circuit
%
% Read PSPICE results using load function
load 'ex6_3ps.dat' -ascii;
```

```
time = ex6_3ps(:,1);
idiode = ex6_3ps(:,2);
plot(time, idiode),    % plot of diode current
xlabel('Time, s')
ylabel('Diode Current, A')
title('Diode Current as a Function of Time')
i_peak = max(idiode);    % peak current
i_ave = mean(idiode);    % average current
% Function trapz is used to integrate the current
charge = trapz(time, idiode);
% Print out the results
fprintf ('Peak Current is % 10.5e A\n', i_peak)
fprintf('Average current is % 10.5e A\n', i_ave)
fprintf('Total Charge is %10.5e C\n', charge)
```

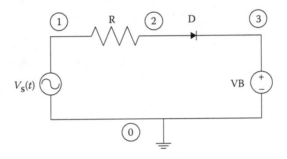

FIGURE 6.5
Battery charging circuit.

The results from MATLAB are:

Peak current is 6.32700e−002 A;

Average current is 8.66958e−003 A; and

Total charge is 4.37814e−004 C.

The plot of the current versus time is shown in Figure 6.6.

A full-wave rectifier that uses a center-tapped transformer is shown in Figure 6.7. When $v_s(t)$ is positive, the diode D1 conducts, but diode D2 is off, and the output voltage is:

$$v_0(t) = v_s(t) - v_D \tag{6.9}$$

where v_D is a voltage drop across a diode.

When $v_s(t)$ is negative, diode D1 is cut-off but diode D2 conducts. The current flowing through the load R enters the latter through node A. The output voltage is:

$$v_s(t) = |v_s(t)| - v_D \tag{6.10}$$

TABLE 6.3

Current Flowing Through a Diode

Time, s	Current, A
0.000E+00	−1.210E−11
1.000E−03	4.201E−09
2.000E−03	6.160E−09
3.000E−03	3.385E−02
4.000E−03	6.148E−02
5.000E−03	4.723E−02
6.000E−03	2.060E−03
7.000E−03	−4.583E−09
8.000E−03	−3.606E−09
9.000E−03	−2.839E−09
1.000E−02	−2.070E−09
1.100E−02	−1.268E−09
1.200E−02	−4.366E−10
1.300E−02	3.940E−10
1.400E−02	1.231E−09
1.500E−02	2.041E−09
1.600E−02	2.820E−09
1.700E−02	3.601E−09

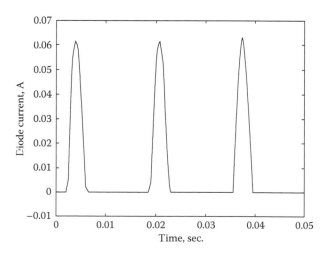

FIGURE 6.6
Diode current.

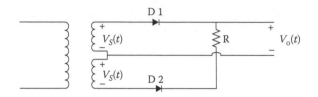

FIGURE 6.7
Full-wave rectifier with center-tapped transformer.

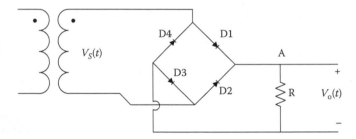

FIGURE 6.8
Bridge rectifier.

A full-wave rectifier that does not require a center-tapped transformer is the bridge rectifier. It is shown in Figure 6.8.

When $v_s(t)$ is negative, the diode D2 and D4 conduct, but diodes D1 and D2 do not conduct. The current entering load resistance R enters it through node A. The output voltage is:

$$v_0(t) = |v_s(t)| - 2v_D \qquad (6.11)$$

When $v_s(t)$ is positive, the diodes D1 and D3 conduct, but diodes D2 and D4 do not conduct. The current entering the load resistance, R, enters the latter through node A. The output voltage is given by Equation (6.11).

Connecting a capacitor across the load can smooth the output voltage of a full-wave rectifier. The resulting circuit is shown in Figure 6.9. The following example explores some characteristics of the smoothing circuit.

Example 6.4: Characteristics of Bridge Rectifier with Smoothing Filter

For Figure 6.9, $v_s(t) = 120\sqrt{2}\sin(2\pi60t)$, C = 100 μF, RS = 1 Ω, LP = 2 H, LS = 22 mH, diodes D1, D2, D3, and D4 are D1N4150. In addition, the coefficient of coupling of the transformer is 0.999. (a) Determine average output voltage and rms value of ripple voltage as RL takes the following values: 10 KΩ, 30 KΩ, 50 KΩ, 70 KΩ, 90 KΩ; (b) plot the (i) the average output voltage, and (ii) rms value of output voltage as a function of RL.

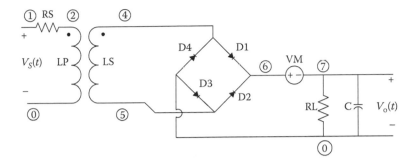

FIGURE 6.9
Full-wave rectifier with capacitor smoothing filter.

Solution

The PSPICE command .STEP is used to vary the element values of resistance RL. The diode current and the output voltage are obtained from PSPICE simulation. Further analysis of the voltage and current are done using MATLAB®.

PSPICE Program

```
BRIDGE RECTIFIER
* SINUSOIDAL TRANSIENT INPUT
VS    1    0     SIN(0 169V 60HZ)
RS    1    2    1
* TRANSFORMER SECTION
LP    2    0     2H
LS    4    5     22MH
KFMR  LP   LS    0.999
*RECTIFIER DIODES
D1    4    6     D1N4150
D2    5    6     D1N4150
D3    0    5     D1N4150
D4    0    4     D1N4150
.MODEL    D1N4150    D(IS=10E-15 RS=1.0 CJO=1.3P TT=12N
BV=70 IBV=0.1P)
*DIODE CURRENT MONITOR
VM    6    7     DC    0
RL    7    0     RMOD  1
C     7    0     100E-6
.MODEL RMOD RES(R = 1)
* ANALYSIS REQUESTS
.TRAN     0.2MS    100MS
.STEP     RES      RMOD(R)     10K     90K     20K
.PRINT    TRAN     V(7)
.PROBE  V(7)
.END
```

Partial PSPICE results showing the output voltage for RL = 10 KΩ and RL = 90 KΩ are shown in Tables 6.4 and 6.5, respectively. The complete results for RL = 10 KΩ,

TABLE 6.4

Output Voltage for Load Resistance of 10 K

Time, s	Output Voltage for RL = 10 KΩ, V
5.000E−03	1.620E+01
1.000E−02	1.612E+01
1.500E−02	1.604E+01
2.000E−02	1.596E+01
2.500E−02	1.611E+01
3.000E−02	1.622E+01
3.500E−02	1.614E+01
4.000E−02	1.616E+01
4.500E−02	1.608E+01
5.000E−02	1.616E+01
5.500E−02	1.617E+01
6.000E−02	1.609E+01
6.500E−02	1.620E+01
7.000E−02	1.611E+01

TABLE 6.5

Output Voltage for Load Resistance of 90 K

Time, s	Output Voltage for RL = 90 KΩ, V
5.000E−03	1.621E+01
1.000E−02	1.620E+01
1.500E−02	1.619E+01
2.000E−02	1.619E+01
2.500E−02	1.618E+01
3.000E−02	1.617E+01
3.500E−02	1.616E+01
4.000E−02	1.615E+01
4.500E−02	1.615E+01
5.000E−02	1.614E+01
5.500E−02	1.621E+01
6.000E−02	1.620E+01
6.500E−02	1.620E+01
7.000E−02	1.619E+01

30 KΩ, 50 KΩ, 70 KΩ, and 90 KΩ can be found in files ex6_4aps.dat, ex6_4bps.
dat, ex6_4cps.dat, ex6_4dps.dat, and ex6_4eps.dat, respectively. In the data analysis, the output voltage stabilizes after 5 ms. The output voltages for times less
than 5 ms have been deleted in calculating the ripple voltage parameters since the
output voltage stabilizes after 5 ms. The MATLAB program for the analysis of the
PSPICE results is as follows:

MATLAB Script

```
% Read PSPICE results using textread
% Load resistors
load 'ex6_4aps.dat' -ascii;
load 'ex6_4bps.dat' -ascii;
load 'ex6_4cps.dat' -ascii;
load 'ex6_4dps.dat' -ascii;
load 'ex6_4eps.dat' -ascii;
t2 = ex6_4aps(:,1);
v10k = ex6_4aps(:,2);
v30k = ex6_4bps(:,2);
v50k = ex6_4cps(:,2);
v70k = ex6_4dps(:,2);
v90k = ex6_4eps(:,2);
rl(1) = 10e3; rl(2) = 30e3; rl(3) = 50e3; rl(4) = 70e3;
rl(5) = 90e3;
%
% Average DC Voltage calculation
v_ave(1) = mean (v10k);
v_ave(2) = mean (v30k);
v_ave(3) = mean (v50k);
v_ave(4) = mean (v70k);
v_ave (5) = mean(v90k);
%
% RMS voltage calculation
n = length(v10k)
for i = 1: n
  V1diff(i) = (v10k(i)-v_ave(1))^2;  % ripple voltage squared
  V2diff(i) = (v30k(i)-v_ave(2))^2;  % ripple voltage squared
  V3diff(i) = (v50k(i)-v_ave(3))^2;  % ripple voltage squared
  V4diff(i) = (v70k(i)-v_ave(4))^2;  % ripple voltage squared
  V5diff(i) = (v90k(i)-v_ave(5))^2;  % ripple voltage squared
end
%
% Numerical Integration
Vint1 = trapz(t2, V1diff);
Vint2 = trapz(t2, V2diff);
Vint3 = trapz(t2, V3diff);
Vint4 = trapz(t2, V4diff);
Vint5 = trapz(t2, V5diff);
%
tup = t2(n); % Upper Limit of integration
v_rms(1) = sqrt(Vint1/tup);
v_rms(2) = sqrt(Vint2/tup);
v_rms(3) = sqrt(Vint3/tup);
v_rms(4) = sqrt(Vint4/tup);
v_rms(5) = sqrt(Vint5/tup);
%
% plot the average voltage
subplot(211)
```

```
plot(rl,v_ave)
ylabel('Average Voltage, V')
title('Average Voltage as a Function of Load Resistance')
% Plot rms voltage vs. RL
subplot(212)
plot(rl, v_rms)
title('Rms Voltage as a Function of Load Resistance')
xlabel('Load Resistance')
ylabel('RMS Value')
```

Figure 6.10 shows the MATLAB plots for the rms voltage and average diode current. Figure 6.10 shows that, in general, as the load resistance increases, the average voltage of the output increases and the rms value of the output ripple voltage decreases.

6.3 Schematic Capture of Diode Circuits

The ORCAD CAPTURE can be used to draw and simulate diode circuits. You start the ORCAD schematic using the steps outlined in Box 1.1. You draw the diode circuit using the steps outlined in Box 1.2. In the latter box, you choose the diode by selecting the diode part. In the Student version of the ORCAD CAPTURE, the diode "DBREAK" can be selected from the

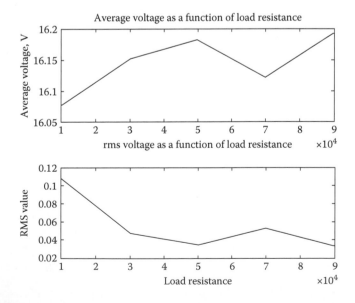

FIGURE 6.10
Average voltage and rms voltage as a function of load resistance.

**BOX 6.1   SEQUENCE OF STEPS FOR SIMULATING
DIODE CIRCUITS**

- Use the steps in Box 1.1 to start ORCAD schematic.
- Use the steps in Box 1.2 to draw the circuit using ORCAD schematic.
- In the steps in Box 1.2, you can select diode part. In the Student version of the ORCAD Schematic package, the diode "DBREAK" can be selected from the "BREAKOUT' library or other diodes can be selected from the "EVAL" library.
- Use Boxes 1.3, 1.4, 1.5, and 1.6 to perform DC, DC Sweep, Transient, and AC Analysis, respectively.

'BREAKOUT" library. Other diodes can be selected from the "EVAL" library. DC, DC Sweep, Transient analysis, and AC analysis can be performed by using the steps outlined in Boxes 1.3, 1.4, 1.5, and 1.6, respectively. Box 6.1 shows the steps needed to perform the analysis of diode circuits.

Example 6.5: Half-Wave Rectifier

In Figure 6.4, if VS is a transient sinusoid with voltage amplitude of 10 V, frequency of 1000 Hz and an average value of 0 V. Use ORCAD Schematic to determine the voltage across the resistor, if R = 1 K. Assume that the diode is D1N914.

Solution

Figure 6.4 is drawn using the schematic capture. VOFF, VAMPL and FREQ are parameters of the transient sinusoid in ORCAD. Transient analysis was performed. Figure 6.11 shows the schematic of the circuit. In addition, Figure 6.12 shows the output voltage.

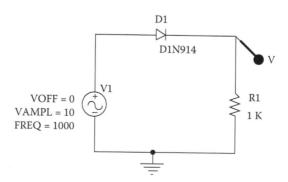

FIGURE 6.11
Half-wave rectifier.

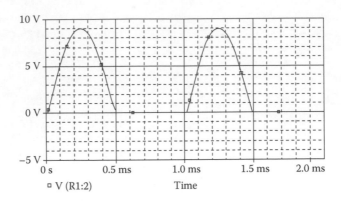

FIGURE 6.12
Output voltage of the half-wave rectifier.

6.4 Zener Diode Voltage Regulator

Zener diode is a pn-junction with a controlled breakdown voltage. The current–voltage characteristic of a zener diode is shown in Figure 6.13. I_{ZK} is the minimum current needed for the breakdown of the zener diode. I_{ZM} is the maximum current that can flow through the zener diode without destroying the latter. It is obtained by:

$$I_{ZM} = \frac{P_Z}{V_Z} \qquad (6.12)$$

where
P_Z is the zener power dissipation; and
V_Z is the zener breakdown voltage.

The incremental resistance or the dynamic resistance of the zener diode at an operating point is specified by:

$$r_Z = \frac{\Delta V_Z}{\Delta I_Z} \qquad (6.13)$$

The following example determines the dynamic resistance of a zener diode.

Example 6.6: Zener Diode Resistance

The circuit shown in Figure 6.14 can be used to obtain the dynamic resistance of a zener diode. Assuming that D1 is D1N4742, RS = 2 Ω, RL = 100 Ω, and $12.2 \leq VS \leq 13.2$, determine the dynamic resistance as a function of diode voltage.

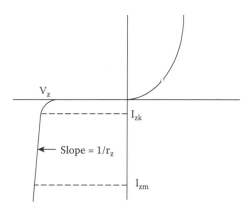

FIGURE 6.13
I–V characteristics of a zener diode.

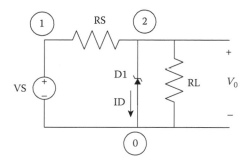

FIGURE 6.14
Zener diode circuit.

Solution

PSPICE will be used to obtain zener current and voltage as the input voltage is varied. MATLAB® is used to obtain the dynamic resistance at various operating points of the zener.

PSPICE Program

```
DYNAMIC RESISTANCE OF ZENER DIODE
.OPTIONS RELTOL=1.0E-08
.OPTIONS NUMDGT=6
VS      1     0        DC      12V
RS      1     2        2
D1      0     2        D1N4742
RL      2     0        100
.MODEL D1N4742 D(IS=0.05UA RS=9 BV=12 IBV=5UA)
.DC     VS    12.2     13.2   0.05
.PRINT        DC       V(0,2)   I(D1)
.END
```

TABLE 6.6

Current Versus Voltage of a
Zener Diode

Voltage, V	Current, A
−1.19608E+01	−1.14718E−06
−1.20098E+01	−7.33186E−06
−1.20587E+01	−4.76843E−05
−1.21073E+01	−2.86514E−04
−1.21544E+01	−1.26124E−03
−1.21992E+01	−3.39609E−03
−1.22424E+01	−6.38033E−03
−1.22846E+01	−9.83217E−03
−1.23264E+01	−1.35480E−02
−1.24090E+01	−2.14124E−02
−1.24501E+01	−2.54748E−02
−1.24910E+01	−2.95934E−02
−1.25319E+01	−3.37552E−02
−1.25726E+01	−3.79510E−02
−1.26134E+01	−4.21743E−02
−1.26541E+01	−4.64202E−02
−1.26947E+01	−5.06850E−02
−1.27354E+01	−5.49660E−02
−1.27760E+01	−5.92607E−02
−1.28165E+01	−6.35675E−02

PSPICE partial results are shown in Table 6.6. The complete results can be found in file ex6_5ps.dat.

MATLAB Script

```
% Dynamic resistance of Zener diode
%
% Read the PSPICE results
load 'ex6_5ps.dat' -ascii;
vd = ex6_5ps(:,2);
id = ex6_5ps(:,3);
n = length(vd); % number of data points
m = n-2;    %number of dynamic resistances to calculate
for i = 1: m
    vpt(i) = vd(i + 1);
    rd(i) = -(vd(i + 2) - vd(i))/(id(i + 2)-id(i));
end
% Plot the dynamic resistance
plot (vpt, rd,'ob',vpt,rd)
title('Dynamic Resistance of a Zener Diode')
xlabel('Voltage, V')
ylabel('Dynamic Resistance, Ohms')
```

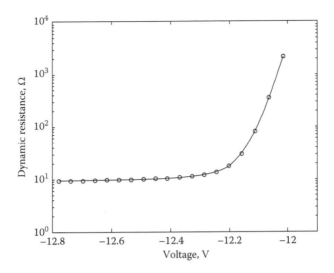

FIGURE 6.15
Dynamic resistance of a zener diode.

The MATLAB plot is shown in Figure 6.15. From Figure 6.15, it seems the dynamic resistance is small at the breakdown region of the zener diode.

The design of voltage reference circuits is one of the applications of a zener diode. A zener diode shunt voltage regulator is shown in Figure 6.14. The circuit, properly designed, will provide a nearly constant output voltage, V_0, which is nearly constant. When the source voltage is greater that the zener breakdown voltage, the zener will breakdown and the output voltage will be equal to the zener breakdown voltage.

The output voltage will change slightly with the variation of the load resistance. This is termed the output voltage regulation. The following example illustrates the change of output voltage as the load resistance is changed.

Example 6.7: Voltage Regulation of Zener Diode Voltage Regulator

In Figure 6.14, VS = 18 V and RS = 400 Ω Find the output voltage as RL varies from 1 K to 51 KΩ. The zener diode is D1N4742.

Solution

PSPICE is used to obtain the output voltage for the various values of load resistance. MATLAB® is used to obtain the plot of output voltage as a function of the load resistance.

PSPICE Program

```
VOLTAGE REGULATOR CIRCUIT
.OPTIONS RELTOL=1.0E-08
.OPTIONS NUMDGT=5
```

```
VS      1     0     DC      18V
RS      1     2     400
DZENER        0     2       D1N4742
.MODEL D1N4742 D(IS=0.05UA   RS=9  BV=12  IBV=5UA)
RL      2     0     RMOD    1
.MODEL    RMOD    RES(R=1)
.STEP   RES     RMOD(R)   1K   51K   5K
*ANALYSIS TO BE DONE
.DC    VS    18    18    1
.PRINT        DC    V(2)
.END
```

PSPICE results are shown in Table 6.7. The PSPICE results are also stored in file ex6_6ps.dat. The MATLAB script for plotting the PSPICE results is as follows:

MATLAB Script

```
% Voltage Regulation
% Plot of Output voltage versus load resistance
% Input the PSPICE results
load 'ex6_6ps.dat' -ascii;
rl = ex6_6ps(:,1);
v = ex6_6ps(:,2);
% Plot rl versus v
plot(rl, v, 'b', rl, v, 'ob')
xlabel('Load Resistance, Ohms')
ylabel('Output Voltage, V')
title('Output Voltage as a Function of Load Resistance')
```

The voltage regulation plot is shown in Figure 6.16.

As the load resistance increases, the output voltage becomes almost constant.

TABLE 6.7

Output Voltage versus Load Resistance

Load Resistance RL, Ω	Output Voltage, V
1.0000E+03	1.2181E+01
6.0000E+03	1.2311E+01
11.0000E+03	1.2321E+01
16.0000E+03	1.2325E+01
21.0000E+03	1.2327E+01
26.0000E+03	1.2328E+01
31.0000E+03	1.2329E+01
36.0000E+03	1.2329E+01
41.0000E+03	1.2330E+01
46.0000E+03	1.2330E+01
51.0000E+03	1.2331E+01

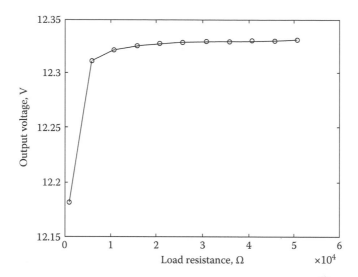

FIGURE 6.16
Load resistance versus output voltage.

Example 6.8: 3-D Plot of Voltage Regulation

In Figure 6.14, the load resistance is varied from 200 Ω to 2000 Ω and the source voltage varies from 4 V to 24 V, while RS is kept constant at 150 Ω. Obtain the output voltage with respect to both the load resistance and the source voltage. Assume that the zener diode is D1N754.

Solution

PSPICE is used to obtain the output voltage as both the source voltage and load resistance is changed.

PSPICE Program

```
VOLTAGE REGULATOR CIRCUIT
VS    1    0      DC    18V
RS    1    2      150
DZENER  0   2      D1N754
.MODEL D1N754 D(IS=880.5E-18 N=1 RS=0.25 IKF=0 XTI=3 EG=1.11
+ CJO=175P M=0.5516 VJ=0.75 FC=0.5 ISR=1.859N NR=2 BV=6.863
+ IBV=0.2723 TT=1.443M)
RL    2    0      RMOD  1
.MODEL  RMOD      RES(R=1)
.STEP  RES   RMOD(R)    0.2E3    2.0E3    0.2E3
* ANALYSIS TO BE DONE
.DC   VS    4    24    1
.PRINT  DC   V(2)
.END
```

TABLE 6.8

Output Voltage as a Function of Load Resistance and
Input Voltage

Source Voltage, V	Load Resistance, Ω	Output Voltage, V
4.000E+00	200	2.286E+00
8.000E+00	200	4.571E+00
1.200E+01	200	6.729E+00
1.600E+01	200	6.810E+00
2.000E+01	200	6.834E+00
2.400E+01	200	6.851E+00
4.000E+00	400	2.909E+00
8.000E+00	400	5.818E+00
1.200E+01	400	6.797E+00
1.600E+01	400	6.827E+00
2.000E+01	400	6.846E+00
2.400E+01	400	6.861E+00
4.000E+00	600	3.200E+00
8.000E+00	600	6.400E+00
1.200E+01	600	6.805E+00
1.600E+01	600	6.831E+00
2.000E+01	600	6.849E+00
2.400E+01	600	6.864E+00

Partial results from PSPICE simulation are shown in Table 6.8. The complete results
can be found in the file ex6_7ps.dat.

MATLAB® is used to do the 3-D plot.

MATLAB Script

```
% 3-D plot of output voltage as a function
% of load resistance and input voltage
%
load 'ex6_7ps.dat' -ascii;
vs = ex6_7ps(:,1);
rl = ex6_7ps(:,2);
vo = ex6_7ps(:,3);
% Do 3-D plot
plot3(vs, rl, vo,'r');
% axis square
grid on
title ('Output Voltage as a Function of Load and Source
Voltage')
xlabel('Input Voltage, V')
ylabel('Load Resistance, Ohms')
zlabel(,Voltage across zener diode,V')
```

The 3-D plot is shown in Figure 6.17.

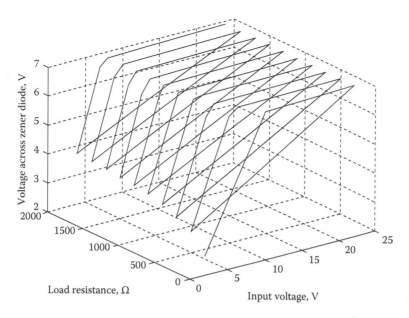

FIGURE 6.17
Voltage versus source voltage and load resistance.

6.5 Peak Detector

Peak detector is a circuit that can be used to detect the peak value of an input signal. Peak detector can also be used as a demodulator to detect the audio signal in an amplitude modulated (AM) signal. A simple peak detector circuit is shown in Figure 6.18.

The peak detector circuit is simply a half-wave rectifier circuit with a capacitor connected across the load resistor. The operation of the circuit will be described with the assumption that the source voltage is a sinusoidal voltage, $V_m \sin(2\pi f_0 t)$, with amplitude greater than 0.6 volts. During the first quarter-cycle, the input voltage will increase and the capacitor will be charged to the input voltage. At time $t = 1/4f_0$, where f_0 is the frequency of the sinusoidal input, the input voltage reaches its maximum value of V_m and the capacitor will be charged to that maximum of V_m.

When time $t = 1/4f_0$, the input starts to decrease, the diode D1 will discharge through the resistance R. If we define $t_1 = 1/4f_0$, the time when the capacitor is charged to the maximum value of the input voltage, the discharge of the capacitor is given as:

$$v_0(t) = V_m e^{-(t-t_1)/RC}. \tag{6.14}$$

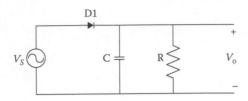

FIGURE 6.18
Peak detector circuit.

The time constant, RC, should be properly selected, to allow the output voltage to approximately represent the peak input signal, within a reasonable error. The following example demonstrates the effects of the time constant.

Example 6.9: Demodulation of AM Wave Using Peak Detection

The peak detector circuit, shown in Figure 6.18, is used to detect an amplitude-modulated wave given as:

$$v_S(t) = 10[1 + 0.5\cos(2\pi f_m t)]\cos(2\pi f_c t) \qquad (6.15)$$

where
f_c = carrier frequency in Hz; and
f_m = modulating frequency in Hz.

If $f_c = 0.2$ MHz, $f_m = 15$ KHz, $C = 20$ nF, RM = 100 Ω, and diode D1 is D1N916, determine the mean square error between the envelope obtained for RL = 1 K and that obtained for the following values of RL: (a) 3 KΩ, (b) 5 KΩ, (c) 7 KΩ, (d) 9 KΩ, and (e) 11 KΩ.

Solution

The modulated wave can be expressed as:

$$v_s(t) = 10\cos(2\pi f_c t) + 5\cos(2\pi f_m t)\cos(2\pi f_c t)$$
$$= 10\cos(2\pi f_c t) + 2.5\cos[2\pi(f_c + f_m)t] + 2.5\cos[2\pi(f_c - f_m)t] \qquad (6.16)$$

PSPICE program uses sine functions only, so we convert the cosine into sine terms using the trigonometric identity:

$$\sin(\omega t + 90°) = \cos(\omega t) \qquad (6.17)$$

Equation 6.16 can be rewritten as:

$$v_s(t) = 10\sin(2\pi f_c t + 90°)$$
$$+ 2.5\sin[2\pi(f_c + f_m)t + 90°] + 2.5\sin[2\pi(f_c - f_m)t + 90°] \qquad (6.18)$$

The frequency content of the modulated wave is:

$$f_c = 200 \text{ KHz}$$

$$f_c + f_m = 215 \text{ KHz}$$

$$f_c - f_m = 185 \text{ KHz}.$$

The demodulator circuit for PSPICE simulation is shown in Figure 6.19.

PSPICE Program

```
AM DEMODULATOR
VS1    1      0     SIN(0  10  200KHZ  0  90)
VS2           2     1     SIN(0  2.5  215KHZ  0  90)
VS3           3     2     SIN(0  2.5  185KHZ   0  90)
D1            3     4     D1N916
.MODEL D1N916 D(IS=0.1P RS=8 CJO=1P TT=12N BV=100 IBV=0.1P)
C      4      0     20E-9
RL     4      0     RMOD 1
.MODEL     RMOD    RES(R=1)
.STEP    RES    RMOD(R)  1K    11K    2K
VS4    5      0     SIN(0  10  200KHZ  0  90)
VS5    6      5     SIN(0  2.5  215KHZ 0  90)
VS6    7      6     SIN(0  2.5  185KHZ  0  90)
RM     7      0     100
.TRAN    2US   150US
.PRINT  TRAN   V(4)    V(7)
.PROBE  V(4)   V(7)
.END
```

Figure 6.20 shows the modulated wave and the envelope for RL of 1 KΩ. PSPICE partial results for RL = 1 KΩ and 11 KΩ are shown in Tables 6.9 and 6.10, respectively. The complete results for RL = 1 KΩ, 3 KΩ, 5 KΩ, 7 KΩ, 9 KΩ, and 11 KΩ can be found in files ex6_8aps.dat, ex6_8bps.dat, ex6_8cps.dat, ex6_8dps.dat, ex6_8eps.dat, and ex6_8fps.dat, respectively.

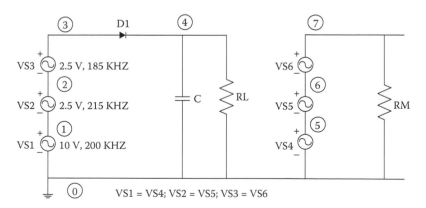

FIGURE 6.19
Demodulator circuit.

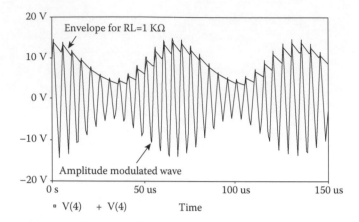

FIGURE 6.20
Amplitude modulated wave and its envelope for load resistance of 1 KΩ.

TABLE 6.9

Output Voltage for Load
Resistance of 1 KΩ

Time, s	Output Voltage for RL = 1 KΩ
1.000E−05	1.100E+01
2.000E−05	7.645E+00
3.000E−05	4.664E+00
4.000E−05	3.670E+00
5.000E−05	6.168E+00
6.000E−05	9.826E+00
7.000E−05	1.156E+01
8.000E−05	1.001E+01
9.000E−05	6.550E+00
1.000E−04	3.973E+00

MATLAB® is used to determine the mean-squared value between the envelope for RL = 1 K and those of the other values of RL.

MATLAB Script

```
% Demodulator circuit
%
% Read Data from PSPICE simulations
load 'ex6_8aps.dat' -ascii;
load 'ex6_8bps.dat' -ascii;
load 'ex6_8cps.dat' -ascii;
load 'ex6_8dps.dat' -ascii;
load 'ex6_8eps.dat' -ascii;
load 'ex6_8fps.dat' -ascii;
```

```
v1 = ex6_8aps(:,2);
v2 = ex6_8bps(:,2);
v3 = ex6_8cps(:,2);
v4 = ex6_8dps(:,2);
v5 = ex6_8eps(:,2);
v6 = ex6_8fps(:,2);
n = length(v1);   % Number of data points
ms1 = 0;
ms2 = 0;
ms3 = 0;
ms4 = 0;
ms5 = 0;
% Calculate squared error
for i = 1:n
    mse1 = ms1 + (v2(i) - v1(i))^2;
    mse2 = ms2 + (v3(i) - v1(i))^2;
    mse3 = ms3 + (v4(i) - v1(i))^2;
    mse4 = ms4 + (v5(i) - v1(i))^2;
    mse5 = ms5 + (v6(i) - v1(i))^2;
end
% Calculate mean squared error
mse(1) = mse1/n;
mse(2) = mse2/n;
mse(3) = mse3/n;
mse(4) = mse4/n;
mse(5) = mse5/n;
rl = 3e3:2e3:11e3
plot(rl, mse, rl, mse, 'ob')
title('Mean Squared Error as Function of Load Resistance')
xlabel('Load Resistance, Ohms')
ylabel('Mean Squared Error')
```

TABLE 6.10

Output Voltage for Load
Resistance of 11 KΩ

Time, s	Output Voltage for RL = 11 KΩ
1.000E−05	1.351E+01
2.000E−05	1.291E+01
3.000E−05	1.234E+01
4.000E−05	1.179E+01
5.000E−05	1.126E+01
6.000E−05	1.143E+01
7.000E−05	1.380E+01
8.000E−05	1.319E+01
9.000E−05	1.260E+01
1.000E−04	1.204E+01

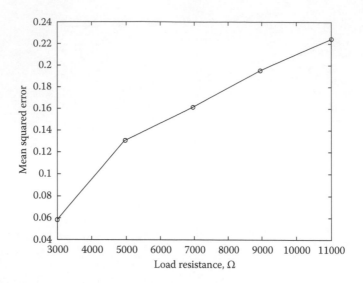

FIGURE 6.21
Mean squared error versus load resistance.

The mean squared error is shown in Figure 6.21. It can be seen from the figure that as the load resistance increases, the mean squared error increases.

6.6 Diode Limiters

A general transfer characteristics of a limiter is given by the following expressions:

$$V_0 = kv_{IN} \qquad V_A \leq v_{IN} \leq V_B$$

$$v_0 = V_H \qquad v_{IN} > V_B$$

$$V_0 = V_L \qquad v_{IN} < V_A \tag{6.19}$$

where
　　k is a constant;
　　V_H and V_L are output high and output low voltages, respectively;
　　v_0 is the output voltage; and
　　v_{IN} is the input voltage.

In Equation 6.19, if v_{IN} exceeds the upper threshold V_B, the output is limited to the voltage V_H. In addition, if the input voltage v_{IN} is less than the lower

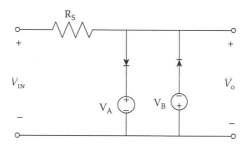

FIGURE 6.22
Double limiter.

threshold voltage V_A, the output is limited to V_L. Equation 6.19 shows the describing relationships for a **double-limiter**, where the output can be limited to two voltages V_H and V_L. If the output voltage is limited to one voltage, either V_H or V_L, then the circuit is a **single-limiter**.

In Equation 6.19, V_H and V_L are constant and are independent of the input voltage v_{IN}. Circuits that have such characteristics are termed **hard limiters.** If V_H and V_L are not constant but are linearly related to the input voltage, then we have **soft limiting.** A limiter circuit is shown in Figure 6.22.

The transfer characteristics of a double limiter are given as:

$$V_0 = V_A + V_D \qquad \text{for} \qquad V_0 \ge V_A + V_D$$

$$V_0 = V_{IN} \qquad \text{for} \qquad -(V_B + V_D) \le V_0 \le V_A + V_D$$

$$V_0 = -(V_B + V_D) \quad \text{for} \qquad V_0 \le -(V_B + V_D) \tag{6.20}$$

where
V_D is the diode drop.

The following example explores a diode limiter circuit.

Example 6.10: Limiter Characteristics

For the precision bipolarity limiter shown in Figure 6.23, VCC = 15 V, VEE = −15 V, R1 = R2 = RA = 10 KΩ. The input signal ranges from −13 V to + 13 V and the diodes are D1N916. (a) Determine (i) the minimum value of output voltage; (ii) the maximum value; and (iii) DC transfer characteristic. (b) What is the proportionality constant of the transfer characteristics where there is no limiting?

Solution
PSPICE is used to do the circuit simulation.

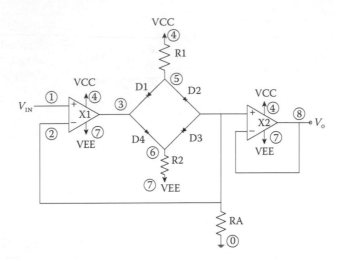

FIGURE 6.23
Precision limiter.

PSPICE Program

```
LIMITER CIRCUIT
* CIRCUIT DESCRIPTION
VIN     1     0     DC     0V;
VCC     4     0     DC     15V;   15V POWER SUPPLY
VEE     7     0     DC     -15V; -15V POWER SUPPLY
X1      1     2     4      7     3        UA741;     UA741 OP-AMP
* +INPUT;-INPUT;+VCC;-VEE;OUTPUT;CONNECTIONS FOR OP AMP UA741
R1      4     5     10K
D1      5     3     D1N916
D2      5     2     D1N916
D3      2     6     D1N916
D4      3     6     D1N916
.MODEL D1N916 D(IS=0.1P RS=8 CJO=1P TT=12N BV=100 IBV=0.1P)
R2      6     7     10K
X2      2     8     4      7     8        UA741;     UA741 OP-AMP
* +INPUT;-INPUT;+VCC;-VEE;OUTPUT;CONNECTIONS FOR OP AMP UA741
RA      2     0     10K
** ANALYSIS TO BE DONE**
* SWEEP THE INPUT VOLTAGE FROM -12V TO +12 V IN 0.2V INCREMENTS
.DC     VIN   -13   13     0.5
** OUTPUT REQUESTED
.PRINT DC V(8)
.PROBE V(8)
.LIB NOM.LIB;
* UA741 OP AMP MODEL IN PSPICE LIBRARY FILE NOM.LIB
.END
```

Partial results from the SPICE simulation are shown in Table 6.11. The complete results can be found in file ex6_9ps.dat.

MATLAB® is used to perform the data analysis and also to plot the transfer characteristics.

TABLE 6.11

Output versus Input Voltages of a Limiter

Input Voltage, V	Output Voltage, V
−1.300E+01	−7.204E+00
−1.100E+01	−7.204E+00
−9.000E+00	−7.204E+00
−7.000E+00	−6.999E+00
−5.000E+00	5.000E∣00
−3.000E+00	−3.000E+00
−1.000E+00	−9.999E−01
1.000E+00	1.000E+00
3.000E+00	3.000E+00
5.000E+00	5.000E+00
7.000E+00	7.000E+00
9.000E+00	7.203E+00
1.100E+01	7.203E+00
1.300E+01	7.203E+00

MATLAB Script

```
% Limiter Circuit
%
% Read data from file
load 'ex6_9ps.dat' -ascii;
vin = ex6_9ps(:,1);
vout = ex6_9ps(:,2);
% Obtain minimum and maximum value of output
vmin = min(vout); % minimum value of output
vmax = max(vout); % maximum value of output
% Obtain proportionality constant of the nonlimiting region
n = length(vin);      % size of data points
% Calculate slopes
for i = 1:n-1
slope(i) = (vout(i + 1) - vout(i))/(vin(i + 1)-vin(i));
end
kprop = max(slope);  % proportionality constant
% Plot the transfer characteristics
plot(vin, vout)
title('Transfer Characteristics of a Limiter')
xlabel('Input Voltage, V')
ylabel('Output Voltage, V')
% Print the results
fprintf('Maximum Output Voltage is %10.4e V\n', vmax)
fprintf('Minimum Output Voltage is %10.4e V\n', vmin)
fprintf('Proportionality Constant is %10.5e V\n', kprop)
```

The transfer characteristics is shown in Figure 6.24.

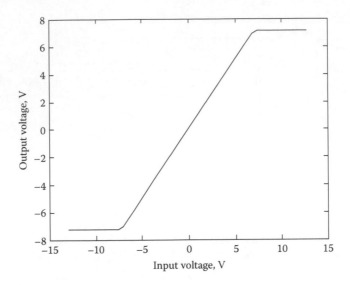

FIGURE 6.24
Transfer characteristics of a limiter.

The following results were obtained from the MATLAB program:

Maximum output voltage is 7.2030e+000 V;

Minimum output voltage is –7.2040e+000 V; and

Proportionality constant is 1.00020e+000 V.

Problems

6.1 A forward-biased diode has the corresponding voltage and current, shown in Table P6.1. (i) Determine the equation of best fit. (ii) For the voltage of 0.64 V, what is the diode current?

TABLE P6.1

Voltage versus Current of a Diode

Forward-Biased Voltage, V	Forward Current, A
0.1	1.33e–13
0.2	1.79e–12
0.3	24.02e–12
0.4	0.321e–9
0.5	4.31e–9
0.6	57.69e–9
0.7	7.72e–7

6.2 For Example 6.2, plot the diode current as a function of temperature. Determine the equation of best fit between the diode current and temperature.

6.3 For diode circuit shown in Figure P6.3, R = 10 K and D1 is D1N916. If the voltage VDC increases from 0.3 V to 2.1 V in vincrements of 0.2 V, determine the diode dynamic resistance and the diode voltage for the various values of the input voltage. Plot the dynamic resistance versus diode voltage.

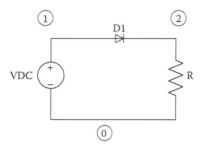

FIGURE P6.3
Diode circuit.

6.4 For the battery charging circuit shown in Figure 6.5, $v_s(t) = 18\sin(120\pi t)$, R = 100 Ω, and VB = 12 V. Determine the conduction angle.

6.5 For the battery charging circuit shown in Figure 6.5, $v_s(t) = 18\sin(120\pi t)$, R = 100 Ω and VB increases from 10.5 to 12.0 V. Plot the average current flowing through the diode as a function of the battery voltage.

6.6 For the half-wave rectifier with smoothing circuit shown in Figure P6.6, diode D1 is D1N4009, LP = 1 H, LS = 100 mH, RS = 5 Ω, and RL = 10 KΩ. (a) If C = 10 μF, plot the diode current with respect to time; (b) If capacitor C takes on the values of 1 μF, 10 μF, 100 μF, and 500 μF,

  i. Plot the rms value of the output voltage with respect to the capacitance values.

  ii. Plot the average diode current with respect to the capacitance values.

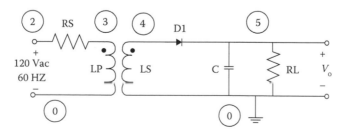

FIGURE P6.6
Rectifier circuit.

6.7   For the full-wave rectifier with the smoothing circuit shown in Figure 6.9, diodes D1, D2, D3, and D4 are D1N914 and C = 10 µF. Determine the peak diode current as RL takes on the following values: 10 KΩ, 30 KΩ, 50 KΩ, 70 KΩ, 90 KΩ, and 110 KΩ. Plot the peak diode current as a function of RL.

6.8   For Figure 6.14, assume that VS = 18 V, D1 is D1N4742, RL = 100 Ω, and RS = 2 Ω. Find the voltage across the diode as the temperature varies from 0°C to 100°C. Plot the diode voltage as a function of temperature.

6.9   In the zener diode shunt voltage regulator shown in Figure 6.14, the source voltage is 20 V, and the zener diode is D1N4742. The load resistance varies from 1 K to 20 KΩ and the source resistance RS is from 20 Ω to 100 Ω. Plot the output voltage as a function of both the load resistance and the source resistance.

6.10 In Figure 6.14, VS varies from 8 V to 20 V and RL = 2 KΩ. The resistance RS varies from 20 Ω to 120 Ω. The zener diode is 1N4742. Plot the output voltage as a function of both the source voltage VS and source resistance RS.

6.11 In the peak detector circuit shown in Figure 6.18, $v_s(t) = 15\sin(360\pi t)V$, C = 0.01 µF, and the diode is D1N4009. As the load resistance takes on the values of 1 KΩ, 3 KΩ, 5 KΩ, 7 KΩ, and 9 KΩ.
   a.   Determine the peak diode currents as a function of load resistance.
   b.   Obtain mean value of the diode current for one complete period of the input signal.

6.12 In Figure 6.23, R1 = R2 = RA = 10 KΩ and $V_{IN}(t) = 10\sin(240\pi t)$ V. X1 and X2 are 741 op amps. If $VCC = |VEE| = V_K$ and V_K is changed from 11 V to 16 V, what will be the minimum and maximum values of the output voltage?

6.13 For the limiter circuit shown in Figure P6.13, VCC = 15 V, VEE = –15 V, R1 = 2 KΩ, R2 = 4 KΩ, and the diodes are D1N754. Determine the peak and minimum values of the output voltage if the sinusoidal voltage of $20\sin(960\pi t)$ is applied at the input. Plot the output voltage. Assume a 741 op amp.

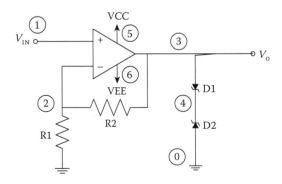

FIGURE P6.13
Limiter circuit.

6.14 In the back-to-back diode limiter circuit shown in Figure P6.14, $R1 = 1 \ K\Omega$ and the diodes D1 and D2 are D1N914. Determine the output voltage if the input signal is sinusoidal wave $V_{IN}(t) = 2\sin(960\pi t)$ V. Calculate the percent distortion for the 2nd, 3rd, and 4th harmonics.

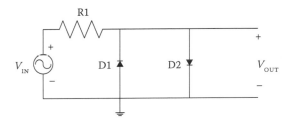

FIGURE P6.14
Back-to-back diode limiter.

6.15 In Figure 6.7, if R is replaced by a combined voltage source of 10 V and a resistor of 100 Ω, determine the following if $v_S(t) = 18\sin(120\pi t)$ V:
 i. the peak current flowing through the resistor,
 ii. the average current flowing through the voltage source.

6.16 The current flowing through a diode at specific instants of time is given in Table P6.16. (a) Plot the current with respect to time. (b) Obtain the average current flowing through the diode.

TABLE P6.16

Current Flowing Through a Diode

Time, s	Current, A
0.000E+00	1.210E−9
5.000E−03	4.201E−07
10.000E−03	6.160E−07
15.000E−03	3.38
20.000E−03	6.148
25.000E−03	4.723
30.000E−03	2.060E−01
35.000E−03	4.583E−07
40.000E−03	3.606E−07
45.000E−03	2.839E−07
5.000E−02	2.070E−07
5.500E−02	−1.268E−07
6.000E−02	−4.366E−08
6.500E−02	3.940E−08
7.000E−02	1.231E−07
7.500E−02	2.041E−07
8.000E−02	2.820E−07
8.500E−02	3.601E−07

6.17   The input and output voltages of a zener voltage regulator of Figure 6.14 are shown in Table P6.17.

If RS = 2 K, RL = 10 K in Figure 6.14, (i) plot the current flowing through the zener diode at each input voltage with respect to the output voltage, (ii) determine the dynamic resistance of the zener.

TABLE P6.17

Input and Output Voltages of Zener Voltage Regulator

Input Voltage VS, V	Output Voltage VOUT, V
12.0	8.756
13.0	8.764
14.0	8.777
15.0	8.783
16.0	8.798
17.0	8.806
18.0	8.905

Bibliography

1. Alexander, Charles K., and Matthew N. O. Sadiku. *Fundamentals of Electric Circuits.* 4th ed. New York: McGraw-Hill, 2009.
2. Attia, J. O. *Electronics and Circuit Analysis Using MATLAB®.* 2nd ed. Boca Raton, FL: CRC Press, 2004.
3. Boyd, Robert R. *Tolerance Analysis of Electronic Circuits Using MATLAB®.* Boca Raton, FL: CRC Press, 1999.
4. Chapman, S. J. *MATLAB® Programming for Engineers.* Tampa, FL: Thompson, 2005.
5. Davis, Timothy A., and K. Sigmor. *MATLAB® Primer.* Boca Raton, FL: Chapman & Hall/CRC, 2005.
6. Distler, R. J. "Monte Carlo Analysis of System Tolerance." *IEEE Transactions on Education* 20 (May 1997): 98–101.
7. Etter, D. M. *Engineering Problem Solving with MATLAB®.* 2nd ed. Upper Saddle River, NJ: Prentice Hall, 1997.
8. Etter, D. M., D. C. Kuncicky, and D. Hull. *Introduction to MATLAB® 6.* Upper Saddle River, NJ: Prentice Hall, 2002.
9. Hamann, J. C, J. W. Pierre, S. F. Legowski, and F. M. Long. "Using Monte Carlo Simulations to Introduce Tolerance Design to Undergraduates." *IEEE Transactions on Education* 42, no. 1 (February 1999): 1–14.
10. Gilat, Amos. *MATLAB®, An Introduction With Applications.* 2nd ed. New York: John Wiley & Sons, 2005.
11. Hahn, Brian D., and Daniel T. Valentine. *Essential MATLAB® for Engineers and Scientists.* 3rd ed. New York and London: Elsevier, 2007.

12. Herniter, Marc E. *Programming in MATLAB®*. Florence, KY: Brooks/Cole Thompson Learning, 2001.
13. Howe, Roger T., and Charles G. Sodini. *Microelectronics, An Integrated Approach*. Upper Saddle River, NJ: Prentice Hall, 1997.
14. Moore, Holly. *MATLAB® for Engineers*. Upper Saddle River, NJ: Pearson Prentice Hall, 2007.
15. Nilsson, James W., and Susan A. Riedel. *Introduction to PSPICE Manual Using ORCAD Release 9.2 to Accompany Electric Circuits*. Upper Saddle River, NJ: Pearson/Prentice Hall, 2005.
16. OrCAD Family Release 9.2. San Jose, CA: Cadence Design Systems, 1986–1999.
17. Rashid, Mohammad H. *Introduction to PSPICE Using OrCAD for Circuits and Electronics*. Upper Saddle River, NJ: Pearson/Prentice Hall, 2004.
18. Sedra, A. S., and K. C. Smith. *Microelectronic Circuits*. 5th ed. Oxford: Oxford University Press, 2004.
19. Spence, Robert, and Randeep S. Soin. *Tolerance Design of Electronic Circuits*. London: Imperial College Press, 1997.
20. Soda, Kenneth J. "Flattening the Learning Curve for ORCAD-CADENCE PSPICE." *Computers in Education Journal* XIV (April–June 2004): 24–36.
21. Svoboda, James A. *PSPICE for Linear Circuits*. 2nd ed. New York: John Wiley & Sons, Inc., 2007.
22. Tobin, Paul. "The Role of PSPICE in the Engineering Teaching Environment." Proceedings of International Conference on Engineering Education, Coimbra, Portugal, September 3–7, 2007.
23. Tobin, Paul. *PSPICE for Circuit Theory and Electronic Devices*. San Jose, CA: Morgan & Claypool Publishers, 2007.
24. Tront, Joseph G. *PSPICE for Basic Circuit Analysis*. New York: McGraw-Hill, 2004.
25. *Using MATLAB®, The Language of Technical Computing, Computation, Visualization, Programming, Version 6*. Natick, MA: MathWorks, Inc., 2000.
26. Yang, Won Y., and Seung C. Lee. *Circuit Systems with MATLAB® and PSPICE*. New York: John Wiley & Sons, 2007.

7

Operational Amplifier

Operational amplifiers (op amps) are highly versatile. They can be used to perform mathematical operations such as addition, subtraction, multiplication, integration, and differentiation. Several electronic circuits, such as amplifiers, filters, oscillators, and flip-flops, use operational amplifiers as an integral element. In this chapter, the properties of op amps will be discussed. The nonideal characteristics will be explored. The analysis of op amp filters and comparator circuits will be done using both PSPICE and MATLAB® programs.

7.1 Inverting and Noninverting Configurations

The op amp, from the signal point of view, is a three terminal device. The ideal op amp has infinite resistance, zero output resistance, zero offset voltage, infinite frequency response, infinite common-mode rejection ratio, and an infinite open-loop gain.

A practical op amp will have large but finite open-loop gain in the range from 10^5 to 10^9 Ω. It also has a very large input resistance from 10^6 to 10^{10} Ω. The output resistance might be in the range of 50 to 125 Ω. The offset voltage is small but finite and the frequency response will deviate considerably from the infinite frequency response of the ideal op amp.

7.1.1 Inverting Configuration

One basic configuration of the op amp is the inverted closed loop configuration, shown in Figure 7.1. It can be shown that:

$$\frac{V_0}{V_{IN}} = -\frac{Z_2}{Z_1}$$

(7.1)

and the input impedance is

$$Z_{IN} = Z_i$$

(7.2)

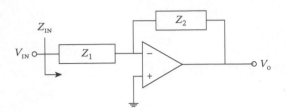

FIGURE 7.1
Inverting configurations.

Case 1: Inverting Amplifier

If $Z_1 = R_1$ and $Z_2 = R_2$, we have an inverting amplifier and the closed loop gain is:

$$\frac{V_0}{V_{IN}} = -\frac{R_2}{R_1} \tag{7.3}$$

Case 2: Miller Integrator

If $Z_1 = R_1$ and $Z_2 = 1/j\omega C$, we obtain an integrator termed Miller Integrator. The closed loop gain of the integrator is:

$$\frac{V_0}{V_{IN}} = -\frac{R_2}{R_1} \tag{7.4}$$

In the time domain, the above expression becomes:

$$V_0(t) = -\frac{1}{CR_1} \int_0^t V_{IN}(t)dt + V_0(0) \tag{7.5}$$

Case 3: Differentiator Circuit

If $Z_1 = 1/j\omega C$ and $Z_2 = R$, we obtain a differentiator circuit. The closed loop gain of a differentiator is:

$$\frac{V_0}{V_{IN}} = -j\omega CR \tag{7.6}$$

In the time domain, the above expression becomes:

$$V_0(t) = -CR\frac{dV_{IN}(t)}{dt} \tag{7.7}$$

The following example explores the behavior of an inverting amplifier.

Example 7.1: DC Transfer Characteristics of Inverting Amplifier

For the inverting amplifier shown in Figure 7.2, VCC = 15 V, VEE = −15 V, R1 = 2 KΩ, and R2 = 5 KΩ.

(i) Determine the maximum and minimum output voltages.
(ii) What is the gain in the region that provides linear amplification?
(iii) Determine the range of input voltage that provides linear amplification.

Solution

PSPICE is used to obtain the transfer characteristics. MATLAB® is then employed to do data processing.

PSPICE Program

```
DC TRANSFER CHARACTERISTICS
VIN   1   0   DC   0.5V
R1    1   2   1E3
R2    2   3   5E3
VCC   5   0   DC   15V;  POWER SUPPLY
VEE   6   0   DC   -15V; POWER SUPPLY
X1    0   2   5   6   3 UA741;  UA741 OP AMP
* +INPUT; -INPUT; +VCC; -VEE; OUTPUT; CONNECTIONS FOR
UA741
** ANALYSIS TO BE DONE
.DC VIN -14 +14 0.5V
.LIB NOM.LIB
* UA741 OP AMP MODEL IN PSPICE LIBRARY FILE NOM.LIB
** OUTPUT
.PRINT DC V(3)
.END
```

Partial results from the PSPICE simulation are shown in Table 7.1. The complete results are in file ex7_1ps.dat.

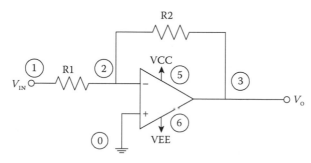

FIGURE 7.2
Inverting amplifier.

TABLE 7.1

Input and Output Voltages of
Inverting Amplifier of Figure 7.2

Voltage Input, V	Output Voltage, V
−1.400E+01	1.461E+01
−5.000E+00	1.461E+01
−3.000E+00	1.460E+01
−1.000E+00	5.000E+00
0.000E+00	5.143E−04
1.000E+00	−4.999E+00
3.000E+00	−1.460E+01
5.000E+00	−1.461E+01
1.400E+01	−1.461E+01

MATLAB is used for processing the input/output data of the amplifier.

MATLAB Script

```
% Analysis of input/output data using MATLAB®
% Read data using load command
load 'ex7_1ps.dat'
vin = ex7_1ps(:,1);
vo = ex7_1ps(:,2);
% Plot transfer characteristics
plot(vin, vo)
xlabel('Input Voltage, V')
ylabel('Output Voltage, V')
title('Transfer Characteristics')
vo_max = max(vo);    % maximum value of output
vo_min = min(vo);    % minimum value of output
% calculation of gain
m = length(vin);
m2 = fix(m/2);
gain = (vo(m2 + 1) - vo(m2 - 1))/(vin(m2 + 1)
- vin(m2 - 1));
% range of input voltage with linear amp
vin_min = vo_min/gain;   % maximum input voltage
vin_max = vo_max/gain;      % minimum input voltage
% print out
fprintf ('Maximum Output Voltage is %10.4eV\n',
vo_max)
fprintf('Minimum Output Voltage is %10.4eV\n', vo_min)
fprintf('Gain is %10.5e\n', gain)
fprintf('Minimum input voltage for Linear Amplification
is %10.5e\n,', vin_max)
fprintf('Maximum input voltage for Linear Amplification
is %10.5e\n,', vin_min)
```

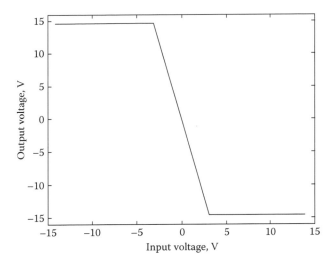

FIGURE 7.3
DC transfer characteristics of an inverting amplifier.

The MATLAB results are:

Maximum output voltage is 1.4610e+001V;
Minimum output voltage is −1.4610e+001V;
Gain is −4.99949e+000;
Minimum input voltage for linear amplification is −2.92230e+000; and
Maximum input voltage for linear amplification is 2.92230e+000.

The transfer characteristics are shown in Figure 7.3.

7.1.2 Noninverting Configuration

Another basic configuration of the op amp is the noninverting configuration. It is shown in Figure 7.4. It can easily be shown that the input and output voltages are related by:

$$\frac{V_0}{V_{IN}} = 1 + \frac{Z_0}{Z_{IN}} \tag{7.8}$$

If $Z_1 = R_1$ and $Z_2 = R_2$, figure 7.4 becomes a voltage follower with gain. The gain is:

$$\frac{V_0}{V_{IN}} = 1 + \frac{R_2}{R_1} \tag{7.9}$$

The noninverting configuration has very high input resistance and the output voltage is in phase with the input voltage. The following example explores the frequency response of a noninverting amplifier.

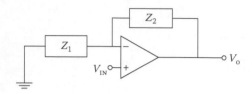

FIGURE 7.4
Noninverting configuration.

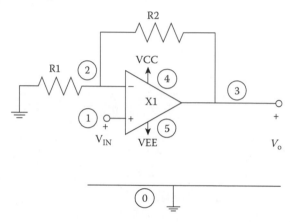

FIGURE 7.5
Noninverting op amp.

Example 7.2: Unity Gain Bandwidth of an Op Amp

For the noninverting amplifier shown in Figure 7.5, VCC = 15 V, VEE = −15 V, R1 = 1 K, and R2 = 9 K. Determine the frequency response of the amplifier. Calculate (a) the 3 dB frequency, (b) unity–gain bandwidth, and (c) gain at mid-band. Assume that X1 is a 741 op amp.

Solution

PSPICE is used to obtain the frequency response. MATLAB® is then employed to calculate the 3-dB frequency, unity-gain bandwidth and the gain at midband.

PSPICE Program

```
FREQUENCY RESPONSE OF NONINVERTING AMP
VIN 1   0   AC  1V  0
R1   2   0   1000
R2   2   3   9000
X1   1   2   4   5   3 UA741;   UA741 OP-AMP
*  +INPUT; -INPUT; +VCC; -VEE; OUTPUT; CONNECTIONS FOR
UA741
VCC 4   0   DC  15V;    15 V POWER SUPPLY
VEE 5   0   DC  -15V;   -15 V POWER SUPPLY
*  ANALYSIS TO BE DONE
```

```
.AC DEC 5 0.1HZ 100MEGHZ
.LIB NOM.LIB;
* UA741 OP AMP MODEL IN PSPICE LIBRARY FILE NOM.LIB
* OUTPUT
.PRINT AC VDB(3)
.END
```

The partial results from the PSPICE simulation is shown in Table 7.2. The complete results can be found in file ex7_2ps.dat.

MATLAB program for analyzing the PSPICE results follows:

MATLAB Script

```
% Frequency Response of a Noninverting Amplifier
load 'ex7_2ps.dat' -ascii;
freq = ex7_2ps(:,1);
gain = ex7_2ps(:,2);
% Plot the frequency response
plot(freq, gain)
xlabel('Frequency, Hz')
ylabel('Gain, dB')
title('Frequency Response of a Noninverting Amplifier')
% calculations
g_mb = gain(1); % midband gain
% Unity gain bandwidth in frequency at which gain is
unity
% Cut-off frequency is frequency at which gain is
approximately
% 3dB less than the midband gain
m = length(freq);   % number of data points
for i= 1:m
 g1(i) = abs(gain(i) - g_mb +3);
 g2(i) = abs(gain(i));
end
% cut-off frequency
[f6, n3dB] = min(g1);
freq_3dB = freq(n3dB);   % 3dB frequency
% Unity Gain Bandwidth
[f7, n0dB] = min(g2);
freq_0dB = freq(n0dB);   % unity gain
% print results
fprintf('Midband Gain is %10.4e dB\n', g_mb)
fprintf('3dB frequency is %10.4e Hz\n', freq_3dB)
fprintf('Unity Gain Bandwidth is %10.4e Hz\n', freq_0dB)
```

The MATLAB results are:

Midband gain is 2.0000e+001 Db;
3dB frequency is 1.0000e+005 Hz; and
Unity gain bandwidth is 1.0000e+006 Hz.

The frequency response is shown in Figure 7.6.

TABLE 7.2

Frequency Response of a
Noninverting Amplifier

Frequency, Hz	Gain, dB
1.000E–01	2.000E+01
3.981E–01	2.000E+01
1.000E+00	2.000E+01
3.981E+00	2.000E+01
1.000E+01	2.000E+01
3.981E+01	2.000E+01
1.000E+02	2.000E+01
3.981E+02	2.000E+01
1.000E+03	2.000E+01
1.000E+04	1.996E+01
3.981E+04	1.942E+01
1.000E+05	1.721E+01
3.981E+05	7.931E+00
1.000E+06	–9.620E–01
3.981E+06	–1.997E+01
1.000E+07	–3.541E+01
3.981E+07	–5.934E+01
1.000E+08	–7.555E+01

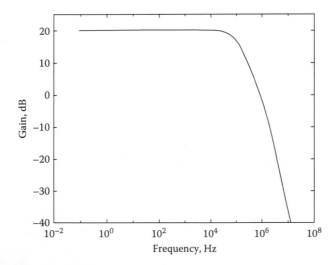

FIGURE 7.6
Frequency response of a noninverting amplifier.

7.2 Slew Rate and Full-Power Bandwidth

Slew rate (SR) is a measure of the maximum possible rate of change of the output voltage of an op amp. Mathematically, it is defined as:

$$SR = \frac{dV_0}{dt}\bigg|_{max} \tag{7.10}$$

Slew rates becomes important when an output signal of an op amp must follow a signal at the input of the op amp that is large in amplitude and rapidly changing with time. If the slew rate is lower than the rate of change of the input signal, then the output will be distorted. However, if the slew rate is higher than the rate of change of the input signal, then no distortion occurs and the input and output of the op amp circuit will have similar wave shapes.

The full power bandwidth, f_m, is the frequency at which a sinusoidal rated output signal begins to show distortion due to slew rate limiting. For a sinusoidal input voltage given by:

$$V_i(t) = V_m \sin(2\pi f_m t) \tag{7.11}$$

If the above signal is applied to a voltage follower with unity gain, then the output rate of change is given as:

$$\frac{dV_0(t)}{dt} = \frac{dV_i(t)}{dt} = 2\pi f_m \cos(2\pi f_m t) \tag{7.12}$$

From Equations 7.10 and 7.12:

$$SR = \frac{dV_0}{dt}\bigg|_{max} = 2\pi f_m V_m \tag{7.13}$$

If the rated output voltage is $V_{0,rated}$, then the slew rate and the full power bandwidth are related by:

$$2\pi f_m V_{0,rated} = SR \tag{7.14}$$

Thus,

$$f_m = \frac{SR}{2\pi V_{0,\ rated}} \tag{7.15}$$

From Equation 7.14, the full-power bandwidth can be traded for output rated voltage. Thus, if the output rated voltage is reduced, the full-power bandwidth increases. The following example explores slew rate and signal distortion.

Example 7.3: Slew Rate and Full-Power Bandwidth

For the voltage follower circuit shown in Figure 7.7, VCC = 15 V and VEE = –15 V. Determine the slew rate if the op amp is a UA741. If the rated output voltage varies from 8 V to 14.5 V, determine the full-power bandwidth. Plot the full-power bandwidth as a function of rated output voltage.

Solution

For the slew rate calculation, a pulse waveform is applied at the input and slope of the rising edge of the output waveform is calculated. Using Figure 7.8, the PSPICE program is used to obtain the output waveform.

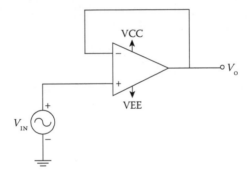

FIGURE 7.7
Voltage follower.

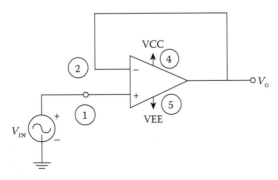

FIGURE 7.8
Slew rate calculation.

PSPICE Program

```
SLEW RATE
* SLEW RATE CALCULATION
VIN1 1  0  PULSE (0 10 0 10NS 10NS 10U 20U)
VCC  4  0  DC       15V
VEE  5  0  DC       -15V
X1   1  2  4    5   2   UA741; UA741 OP AMP
* +INPUT; -INPUT; +VCC; -VEE; OUTPUT; CONNECTIONS FOR
UA741
.LIB NOM.LIB;
* UA741 OP AMP MODEL IN PSPICE LIBRARY FILE NOM.LIB
* ANALYSIS TO BE DONE
.TRAN  0.5U   40U
.PRINT TRAN         V(1) V(2)
.PROBE V(1)         V(2)
.END
```

The partial results of the output voltage obtained from PSPICE are shown in Table 7.3. The results can be found in ex7_3ps.dat.

MATLAB® is used to calculate the slew rate and also to plot the full-power bandwidth as a function of input voltage.

MATLAB Script

```
% Slew rate and full-power bandwidth
% read data
load 'ex7_3ps.dat' -ascii;
t = ex7_3ps(:,1);
vo = ex7_3ps(:,3);
% slew rate calculation
nt = length(t);    % data points
% calculate derivative with MATLAB® function diff
dvo = diff(vo)./diff(t);   % derivative of output with
respect to % time
% find max of the derivative
sr = max(dvo);
vo_rated = 8.0:0.5:14.5;
ko = length(vo_rated);
for i=1:ko
 fm(i)=sr/(2*pi*vo_rated(i));
end
% plot
subplot(211), plot(t,vo)
xlabel('Time,s')
ylabel('Output Voltage')
title('Output Voltage and Full-Power Bandwidth')
subplot(212),plot(vo_rated,fm)
xlabel('Rated Output Voltage, V')
ylabel('Full-power Bandwidth')
fprintf('Slew Rate is %10.5e V/s\n',sr)
```

The output voltage as a function of time was used to calculate the slew rate. Equation 7.15 was used to calculate the full-power bandwidth. The output voltage and the full-power bandwidth plots are shown in Figure 7.9 Slew Rate is 5.22200e+005 V/s.

The following example explores the effects of input voltage amplitude and frequency on the output of an operational amplifier.

TABLE 7.3

Output Voltage versus Time

Time, s	Output Voltage, V
0.000E+00	1.925E−05
2.000E−06	9.948E−01
4.000E−06	2.038E+00
6.000E−06	3.080E+00
8.000E−06	4.123E+00
1.000E−05	5.166E+00
1.200E−05	4.254E+00
1.400E−05	3.237E+00
1.600E−05	2.226E+00
1.800E−05	1.220E+00
2.000E−05	2.189E−01

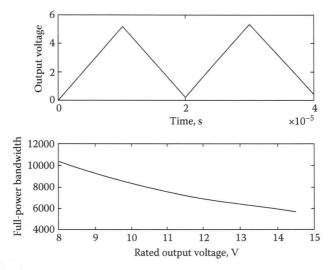

FIGURE 7.9

Ouput voltage and full-power bandwidth.

Example 7.4: 3D Plot of Output Voltage with Respect to Input Voltage and Frequency

For the voltage follower circuit shown in Figure 7.7, the input signal is a sinusoidal signal given as:

$$V_{IN}(t) = V_m \sin(2\pi f t) \tag{7.16}$$

If peak voltage, V_m, varies from 1 to 11 V and the frequency, f, varies from 20 K to 200 KHz, determine the peak output voltage. Plot the peak output voltage as a function of peak input voltage and frequency.

Solution

PSPICE is used to obtain the output voltage with the changes of the amplitude and frequency of the input sinusoid.

PSPICE Program

```
OUTPUT VOLTAGE AS A FUNCTION OF INPUT VOLTAGE AND
FREQUENCY
.PARAM   PEAK = 1.0V
.PARAM FREQ = 20KHz
VIN 1   0   SIN(0 {PEAK} {FREQ})
X2  1  2   4   5  2    UA741; U741 Op Amp
* +INPUT; -INPUT; +VCC; -VEE; OUTPUT; CONNECTIONS FOR
UA741
.LIB NOM.LIB;
* UA741 OP AMP MODEL IN PSPICE LIBRARY FILE NOM.LIB
VCC 4   0   DC   15V
VEE 5   0   DC   -15V
.STEP PARAM FREQ 20KHz 200KHz 20KHz
.TRAN 0.01US 40US
.PRINT TRAN V(2)
.PROBE V(2)
.END
```

The peak value of the input voltage is changed from 1 V to 11 V in increments of 2 V, the peak output voltage is then obtained for each corresponding value of input voltage and frequency. Partial results from PSPICE simulation are shown in Table 7.4. The complete data can be found in file ex7_4ps.dat.

MATLAB® Script

```
% 3D Plot of Output Voltage as a function of input
voltage amplitude
% and frequency.
% Read data
[vin_amp, vin_freq, vout] = textread('ex7_4ps.dat', '%d
%d %f');
% 3D Plot
plot3(vin_amp,vin_freq, vout)
```

```
title('Peak Output Voltage as a function of Amplitude and
Frequency of Input')
xlabel('Amplitude, V')
ylabel('Frequency, Hz')
zlabel('Peak Output Voltage, V')
```

The 3-D plot is shown in Figure 7.10. From Figure 7.10 and Table 7.4, it seems for small input voltage and frequency, the peak output voltage is almost equal to the amplitude of the input signal. However, with high input voltage and frequency, the peak output voltage is considerably lower than the input amplitude. This is due to slew rate limiting.

TABLE 7.4

Output Voltage as a Function of Load Resistance and Input Voltage

Amplitude of Input Voltage, V	Frequency of Input Voltage, KHz	Peak Output Voltage, V
1	20	1.0
1	40	1.0
1	60	1.0
3	20	3.0
3	40	3.0
3	60	2.57
5	20	5.0
5	40	4.235
5	60	3.205
7	20	6.856
7	40	4.235
7	60	3.205

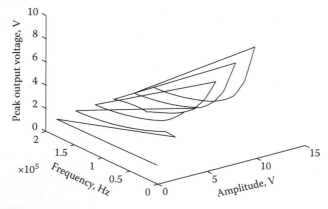

FIGURE 7.10
3D plot of output voltage.

BOX 7.1 SEQUENCE OF STEPS FOR SIMULATING OPERATIONAL AMPLIFIER CIRCUITS

- Use the steps in Box 1.1 to start ORCAD schematic.
- Use the steps in Box 1.2 to draw the circuit using ORCAD schematic.
- In the steps in Box 1.2, select operational amplifier part. In the Student version of the ORCAD Schematic package, select the operational amplifier from the "ANALOG" library.
- The operational amplifier will have model parameters. To set the parameters of the device, select the device and go to Edit > PSPICE Model. Click it to open the PSPICE Model Editor and insert the model for the device. Models of various devices are available on the Internet from device manufacturers' websites.
- Use Boxes 1.3, 1.4, 1.5, and 1.6 to perform DC, DC Sweep, Transient, and AC Analysis, respectively.

7.3 Schematic Capture of Operational Amplifier Circuits

The ORCAD CAPTURE can be used to draw and simulate operational amplifier circuits. Start the ORCAD schematic using the steps outlined in Box 1.1. Draw the operational amplifier circuit using the steps outlined in Box 1.2. In the latter box, choose the operational amplifier by selecting the operational amplifier part. In the Student version of the ORCAD CAPTURE, select the operational amplifier from the ANALOG library. DC, DC Sweep, Transient analysis, and AC analysis can be performed by using the steps outlined in Boxes 1.3, 1.4, 1.5, and 1.6, respectively. Box 7.1 shows the steps needed to perform the analysis of diode circuits.

Example 7.5: An Integrator Circuit

In Figure 7.1, V_{IN} is a square wave with average value of zero, and peak–peak value of 8 V, pulse width of 5 ms, and pulse duration of 10 ms. If Z_1 is a resistor with a value of 10 K, and Z_2 has a capacitance of 0.1 microfarads, determine the output voltage.

Solution

Figure 7.1 is drawn using the schematic capture. Transient analysis was performed. Figure 7.11 shows the schematic of the circuit. In addition, Figure 7.12 shows the output voltage.

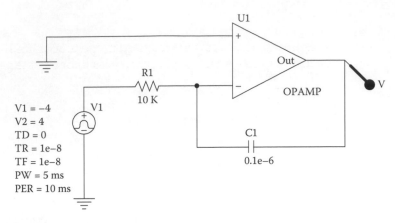

FIGURE 7.11
Operational amplifier circuit.

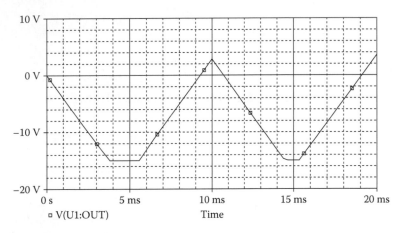

FIGURE 7.12
Output voltage of the Figure 7.11.

7.4 Active Filter Circuits

Electronic filter circuits are circuits that can be used to attenuate particular band(s) of frequency and also pass other band(s) of frequency. The following types of filters will be covered in this section: lowpass, bandpass, highpass, and bandreject. The filters have passband, stopband, and transition band. The order of the filter will determine the transition from the passband to stopband.

7.4.1 Lowpass Filters

Lowpass filter passes low frequencies and attenuates high frequencies. The transfer function of a first order lowpass filter has the general form:

$$H(s) = \frac{k}{s + \omega_0} \tag{7.17}$$

A circuit that can be used to implement a first order lowpass filter is shown in Figure 7.13. The voltage transfer function for Figure 7.13 is:

$$H(s) = \frac{V_0(s)}{V_{IN}(s)} = \frac{k}{1 + sR_1C_1} \tag{7.18}$$

with DC gain, k, given as,

$$k = 1 + \frac{R_F}{R_2} \tag{7.19}$$

and the cut-off frequency, f_0,

$$f_0 = \frac{1}{2\pi R_1 C_1} \tag{7.20}$$

The first-order filter exhibits -20 dB/decade roll-off in the stopband. We shall next consider second order lowpass filter that has -40 dB/decade roll-off. In addition, the second-order lowpass filter can be used as a building block for higher-order filters. The general transfer function for a second order lowpass filter is:

$$H(s) = \frac{k\omega_0^2}{s^2 + (\omega_0/Q)s + \omega_0^2} \tag{7.21}$$

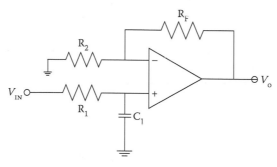

FIGURE 7.13
First order lowpass filter.

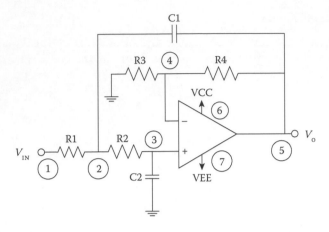

FIGURE 7.14
Sallen–Key lowpass filter.

where
ω_0 is the resonant frequency;
Q is the quality factor or figure of merit; and
k is the dc gain.

The quality factor Q is related to the bandwidth BW and ω_0 by the expression:

$$Q = \frac{\omega_0}{BW} = \frac{\omega_0}{\omega_H - \omega_L} \tag{7.22}$$

where
ω_H = high cutoff frequency, in rad/s; and
ω_L = low cutoff frequency, in rad/s.

A circuit that can be used to realize a second order lowpass filter is the Sallen–Key filter shown in Figure 7.14. The following example explores the characteristics of the Sallen–Key lowpass filter circuit.

Example 7.6: Sallen–Key Lowpass Filter

The Sallen–Key lowpass filter, shown in Figure 7.14, has the following element values: R1 = R2 = 30 KΩ, R3 = 10 KΩ, R4 = 40 KΩ, and C1= C2 = 0.005 µF. (a) Determine low frequency gain, (b) find the cut-off frequency, and (c) plot the frequency response.

Solution

PSPICE is used for circuit simulation.

PSPICE Program

```
*SALLEN-KEY LOWPASS FILTER
VIN 1   0   AC   1V
R1  1   2   30K
R2  2   3   30K
R3  4   0   10K
R4  5   4   40K
C1  2   5   0.005UF
C2  3   0   0.005UF
X1  3   4   6    7    5 UA741; UA741 OP AMP
* +INPUT; -INPUT; +VCC; -VEE; OUTPUT; CONNECTIONS FOR UA741
.LIB NOM.LIB;
* UA741 OP AMP MODEL IN PSPICE LIBRARY FILE NOM.LIB
.AC DEC   10 1HZ 100KHZ
.PRINT     AC VM(5)
.PROBE V(5)
.END
```

The partial results obtained from PSPICE simulation are shown in Table 7.5. The complete output data can be found in file ex7_5ps.dat.

MATLAB® is used for the data analysis.

MATLAB script

```
% Lowpass gain and cut-off frequency
% Read data
load 'ex7_5ps.dat' -ascii;
freq = ex7_5ps(:,1);
vout = ex7_5ps(:,2);
gain_lf = vout(2);   % Low frequency gain
gain_cf = 0.707 * gain_lf;  % gain at cut-off frequency
tol = 1.0e-3;     % tolerance for obtaining cut-off
frequency
i = 2;      % Initialize the counter
% Use while loop to obtain the cut-off frequency
while (vout(i) - gain_cf) > tol
   i = i +1;
end
m=i;
freq_cf = freq(m);      % cut-off frequency
% Print out the results
plot(freq,vout)
xlabel('Frequency, Hz')
ylabel('Gain')
title('Frequency Response of a Lowpass Filter')
fprintf('Low frequency gain is %10.5e\n', gain_lf)
fprintf('Cut-off frequency is %10.5e\n', freq_cf)
```

The results are:

Low frequency gain is 5.00900e+000; and
Cut-off frequency is 7.94300e+002.

TABLE 7.5

Gain versus Frequency of a Lowpass Filter

Frequency, Hz	Gain
1.000E+00	5.009E+00
5.012E+00	5.009E+00
1.000E+01	5.009E+00
5.012E+01	4.998E+00
1.000E+02	4.966E+00
5.012E+02	4.098E+00
1.000E+03	2.644E+00
5.012E+03	2.097E−01
1.000E+04	5.428E−02
5.012E+04	2.176E−03
1.000E+05	9.273E−04

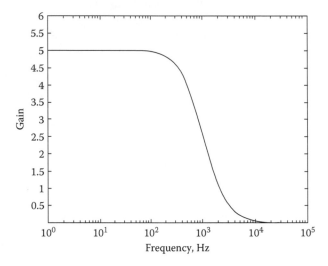

FIGURE 7.15
Frequency response of a lowpass filter.

The frequency response of the lowpass is shown in Figure 7.15.

7.4.2 Highpass Filters

Highpass filter passes high frequencies and attenuates low frequencies. The transfer function of the first order highpass filter has the general form:

$$H(s) = \frac{ks}{s + \omega_0}. \tag{7.23}$$

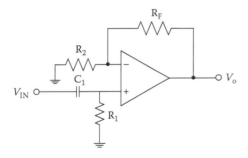

FIGURE 7.16
First order highpass filter.

The circuit, shown in Figure 7.16, can be used to implement the first order highpass filter. It is basically the same as Figure 7.13, except that the positions of R_1 and C_1 in Figure 7.13 have been interchanged. For Figure 7.16, the voltage transfer function is:

$$H(s) = \frac{V_O}{V_{IN}}(s) = \frac{s}{1 + 1/R_1C_1}\left(1 + \frac{R_F}{R_2}\right) \qquad (7.24)$$

where

$$k = \left(1 + \frac{R_F}{R_2}\right) \qquad (7.25)$$

$$= \text{gain at very high frequency}$$

and the cut-off frequency at 3dB gain is,

$$f_0 = \frac{1}{2\pi R_1 C_1}. \qquad (7.26)$$

Although the filter, shown in Figure 7.16, passes all signal frequencies higher than f_0, the high frequency characteristic is limited by the bandwidth of the op amp. The second order highpass filter has the general form:

$$H(s) = \frac{ks^2}{s^2 + (\omega/Q)s + \omega_0^2} \qquad (7.27)$$

where
k is the high frequency gain.

The second-order highpass filter can be obtained from the second-order lowpass filter, shown in Figure 7.17. The latter figure is similar to Figure 7.14 except that the frequency-dominant resistors and capacitors have been interchanged.

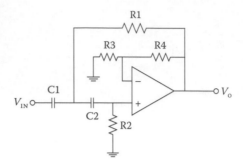

FIGURE 7.17
Sallen–Key highpass filter.

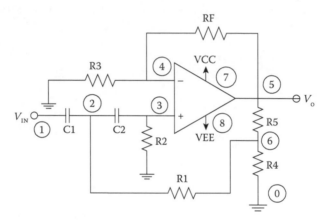

FIGURE 7.18
Modified Sallen–Key highpass filter.

The following example will explore the high-pass filter characteristics.

Example 7.7: Highpass Filter

The Sallen–Key highpass filter, shown in Figure 7.17, may be modified to obtain various quality factors by including two resistors R4 and R5 at the output. The modified Sallen–Key highpass filter is shown in Figure 7.18. In the latter figure, assume that VCC = 15 V, VEE = −15 V, C1 = C2 = 0.05 μF, R1 = R2 = R3 = R4 = 600 Ω, and RF = 3000 Ω. R5 takes on the following values: (a) 450 Ω, (b) 900 Ω, and (c) 1350 Ω. What is the quality factor when R5 = 450 Ω? Plot the frequency response.

Solution

PSPICE is used to obtain the frequency response and MATLAB® is employed to obtain the cut-off frequency and to calculate the quality factor.

PSPICE Program

```
* MODIFIED SALLEN-KEY HIGHPASS FILTER
VIN  1    0   AC   0.5V
VCC  7    0   DC   15V
VEE  8    0   DC   -15V
C1   1    2   0.05e-6
C2   2    3   0.05e-6
R1   2    6   600
R2   3    0   600
R3   4    0   600
R4   6    0   600
RF   5    4   3000
X1 3 4 7 8 5 UA741;      UA741 Op Amp
* +INPUT; -INPUT; +VCC; -VEE; OUTPUT; CONNECTIONS FOR
UA741
.LIB NOM.LIB;
* UA741 OP AMP MODEL IN PSPICE LIBRARY FILE NOM.LIB
.PARAM  VAL = 900
R5   5    6   {VAL}
.STEP    PARAM    VAL LIST    450   900   1350
.AC DEC 20    100Hz    100KHz
.PRINT    AC VM(5)
.PROBE V(5)
.END
```

The partial results of PSPICE simulation for R5 = 450 Ω are shown in Table 7.6. The complete output can be found in file ex7_6aps.dat, ex7_6bps.dat, and ex7_6cps. dat for results; for R5 equals 450 Ω, 900 Ω, and 1350 Ω, respectively.

TABLE 7.6

Output Voltage as a Function of Frequency for R5 = 450 Ω

Frequency, Hz	Gain for R5 = 450 Ω (Multiply Entries by 2)
1.000E+02	3.047E−03
3.162E+02	3.062E−02
5.012E+02	7.751E−02
7.079E+02	1.567E−01
1.000E+03	3.211E−01
3.162E+03	5.876E+00
5.012E+03	1.372E+01
7.079E+03	8.259E+00
1.000E+04	6.655E+00
3.162E+04	5.555E+00
5.012E+04	5.382E+00
7.079E+04	5.187E+00
1.000E+05	4.872E+00

MATLAB Script

```
% load pspice results
load 'ex7_6aps.dat' -ascii;
load 'ex7_6bps.dat' -ascii;
load 'ex7_6cps.dat' -ascii;
fre = ex7_6aps(:,1);
g450 = 2*ex7_6aps(:,2);
g900 = 2*ex7_6bps(:,2);
g1350 = 2*ex7_6cps(:,2);
m = length(fre);
tol = 1.0e-4;
% Plot frequency response
plot(fre,g450, fre,g900, fre, g1350)
xlabel('Frequency, Hz')
ylabel('Gain')
title('Frequency Response of a Sallen-Key Highpass
Filter')
%
% Determine quality factor for R% = 450 Ohms
[gmax, kg] = max(g450);   % maximum value of gain
gcf = 0.707 * gmax;       % cut-off frequency
% determine the cut-off frequencies
% low cut-off frequency, index, lcf
% High cut-off frequency, index, hcf
k = kg;   % initalize counter
while (g450(k) - gcf) > tol
    k = k+ 1;
end
hcf = k;
i = kg;
while (g450(i) - gcf) > tol
    i = i- 1;
end
lcf = i;
% Calculate Quality factor
Qfactor = fre(kg)/(fre(hcf) - fre(lcf))
```

The plot of the frequency response is shown in Figure 7.19. The quality factor when R5 is 450 Ω equals to 2.1528.

7.4.3 Bandpass Filters

Bandpass filter passes a band of frequencies and attenuates other bands. The filter has two cut-off frequencies f_L and f_H. We assume that $f_H > f_L$. All signal frequencies lower than f_L or greater than f_H are attenuated. The general form of the transfer function of a bandpass filter is:

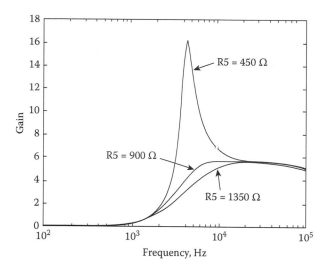

FIGURE 7.19

Frequency response of a modified Sallen–Key highpass filter.

$$H(s) = \frac{k\left(\omega_c/Q\right)s}{s^2 + \left(\omega_c/Q\right)s + \omega_c^2} \tag{7.28}$$

where

k is the passband gain; and

ω_C is the center frequency in rad/s.

The quality factor Q is related to the 3-dB bandwidth and the center frequency by the expression:

$$Q = \frac{\omega_c}{BW} = \frac{f_c}{f_H - f_L}. \tag{7.29}$$

Bandpass filters with $Q \leq 10$ are classified as wide bandpass. On the other hand, bandpass filters with $Q > 10$ are considered narrow bandpass.

Wideband pass filters may be implemented by cascading lowpass and highpass filters. The order of the bandpass filters is the sum of the highpass and lowpass sections. The advantages of this arrangement are that the fall-off, bandwidth, and midband gain can be set independently.

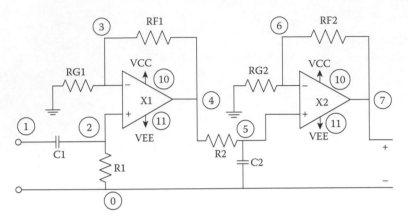

FIGURE 7.20
2nd order wideband pass filter.

Figure 7.20 shows the wideband pass filter, built using first order highpass and first order lowpass filters. The magnitude of the voltage gain is the product of the voltage gains of both the highpass and lowpass filters.

The following example illustrates the characteristics of the wideband filter.

Example 7.8: Second Order Wideband Pass Filter

For the bandpass filter shown in Figure 7.20, assume that op amps X1 and X2 are 741 op amps. If RG1 = RG2 = 1 KΩ, RF1 = RF2 = 5 KΩ, C1 = 30 nF, R1 = 50 KΩ, R2 = 1000 Ω, and C2 = 15 nF. Determine the (a) bandwidth, (b) low cut-off frequency, (c) high cut-off frequency, and (d) quality factor.

Solution
PSPICE is used to obtain the frequency response data.

PSPICE Program

```
BANDPASS FILTER (HIGHPASS PLUS LOWPASS SECTIONS)
VIN 1  0   AC   0.2V
C1  1  2   30E-9
R1  2  0   50E3
RG1 3  0   1E3
RF1 3  4   5E3
X1  2  3   10   11   4   UA741; UA741 OP AMP
* +INPUT; -INPUT; +VCC; -VEE; OUTPUT; CONNECTIONS FOR
UA741
.LIB NOM.LIB;
* UA741 OP AMP MODEL IN PSPICE LIBRARY FILE NOM.LIB
```

```
R2    4   5    1.0E3
C2    5   0    15E-9
X2    5   6    10    11  7  UA741; UA741 OP AMP
RG2   6   0    1.0E3
RF2   6   7    5.0E3
VCC  10   0    DC    15V
VEE  11   0    DC    -15V
.AC   DEC 40 10HZ      100KHZ
.PRINT  AC  VM(7)
.PROBE  V(7)
.END
```

The partial results of the PSPICE simulation are shown in Table 7.7. The complete output can be found in file ex7_7ps.dat. The low cutoff frequency, bandwidth and quality factor are calculated using MATLAB®.

MATLAB Script

```
% Load pspice results
load 'ex7_7ps.dat' -ascii;
fre = ex7_7ps(:,1);
vout = 5*ex7_7ps(:,2);
m = length(fre);
% [gmax, kg] = max(g450);      % maximum value of gain
[g_md, m2] = max(vout);  % mid-band gain
g_cf = 0.707*g_md;       % gain at cut-off
% Determine the low cut-off frequency index
i = m2;
tol = 1.0e-4;
while (vout(i) - g_cf) > tol;
   i = i - 1;
end
lcf = i;
% Determine the high frequency cut-off index
k = m2;
while (vout(k) - g_cf) > tol;
   k = k + 1;
end
hcf = k;
low_cf = fre(lcf);       % Low cut-off frequency
high_cf = fre(hcf);      % High cut-off frequency
ctr_cf = fre(m2);        % Center frequency
band_wd = high_cf - low_cf;       % Bandwidth
Qfactor = ctr_cf/band_wd;
% Print results
low_cf
high_cf
band_wd
Qfactor
```

TABLE 7.7

Output Voltage versus
Frequency for Bandpass Filter

Frequency, Hz	Gain (Multiply Entries by 5)
1.000E+01	6.755E−01
5.012E+01	3.075E+00
1.000E+02	4.937E+00
1.000E+02	4.937E+00
5.012E+02	7.035E+00
1.000E+03	7.127E+00
5.012E+03	6.507E+00
1.000E+04	5.230E+00
5.012E+04	1.386E+00
1.000E+05	5.785E−01

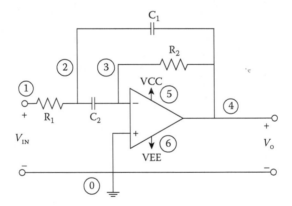

FIGURE 7.21
Multiple-feedback bandpass filter.

The results are shown below:

  Low frequency cut-off is 100 Hz;
  High frequency cut-off is 11220 Hz;
  Bandwidth is 11120 Hz; and
  Quality factor is 0.0952.

  A narrowband pass filter normally has a high Q-value. A circuit that can be used to implement a narrowband filter is a multiple feedback filter. The circuit is shown in Figure 7.21.

The circuit has two feedback paths and one op amp. The circuit can be designed to have a low Q-value and thus manifest a wideband pass filter characteristic. It can be shown that the transfer function of the filter network is:

$$H_{PB} = \frac{V_0(s)}{V_{IN}(s)} = \frac{\left(\frac{-1}{R_1C_1}\right)s}{s^2 + \left(\frac{1}{R_2}\right)\left(\frac{1}{C_1}+\frac{1}{C_2}\right)s + \frac{1}{R_1R_2C_1C_2}} \tag{7.30}$$

$$= \frac{k_{PB}\left(\frac{\omega_C}{Q}\right)s}{s^2 + \left(\frac{\omega_C}{Q}\right)s + \omega_C^2}. \tag{7.31}$$

The center frequency, f_C, is:

$$f_C = \frac{\omega_C}{2\pi} = \frac{1}{2\pi}\frac{1}{\sqrt{R_1R_2C_1C_2}}. \tag{7.32}$$

If C1 = C2 = C, the Q-values is:

$$Q = \frac{1}{2}\sqrt{\frac{R_2}{R_1}}, \tag{7.33}$$

the passband gain, k_{PB} is

$$k_{PB} = \frac{1}{R_1C_1}\left(\frac{Q}{\omega_C}\right). \tag{7.34}$$

The following example explores the characteristics of the multiple feedback bandpass filter.

Example 7.9: Multiple Feedback Narrowband Pass Filter

For the circuit shown in Figure 7.21, the elements have the following values: R1 = 1 KΩ, C1 = C2 = 100 nF. The op amp is UA741. If R2 takes the following values: (a) 25 KΩ, (b) 75 KΩ, and (c) 125 KΩ, what are the quality factor, bandwidth, and center frequencies for the values of R2?

Solution

PSPICE is used to obtain the frequency response data. MATLAB® is used for the data analysis.

PSPICE Program

```
NARROWBAND-PASS FILTER (MULTIPLE-FEEDBACK BANDPASS)
.PARAM R2_VAL = 25K;
VIN 1   0    AC 0.1
R1   1   2    1.0E3
C1   2   4    100.0E-9
C2   2   3    100.0E-9
R2   3   4    {R2_VAL}
.STEP  PARAM  R2_VAL    25K     125K   50K
X1   0   3    5  6    4    UA741
* +INPUT; -INPUT; +VCC; -VEE; OUTPUT; CONNECTIONS FOR
UA741
.LIB NOM.LIB;
* UA741 OP AMP MODEL IN PSPICE LIBRARY FILE NOM.LIB
VCC 5   0    DC   15V
VEE 6   0    DC   -15V
.AC LIN 500 100HZ   1KHZ
.PRINT AC VM(4)
.PROBE V(4)
.END
```

The PSPICE partial results for feedback resistance R2 of 25 KΩ are shown in Table 7.8. The complete PSPICE data can be found in file ex7_8aps.dat, ex7_8bps. dat, and ex7_8cps.dat for R2 equal to (a) 25 KΩ, (b) 75 KΩ, and (c) 125 KΩ, respectively.

The data analysis is done using MATLAB.

TABLE 7.8

Frequency Response for Multiple
Feedback Bandpass Filter for R2 = 25 KΩ

Frequency, Hz	Gain for R2 = 25 KΩ (Multiply Entries by 1000)
1.000E+02	1.726E−01
1.505E+02	2.959E−01
2.010E+02	4.846E−01
2.515E+02	8.059E−01
3.002E+02	1.201E+00
3.507E+02	1.123E+00
4.012E+02	8.111E−01
4.499E+02	6.148E−01
5.004E+02	4.901E−01
5.509E+02	4.089E−01
6.014E+02	3.520E−01
6.501E+02	3.112E−01

MATLAB Script

```
% Load data
load 'ex7_8aps.dat' -ascii;
load 'ex7_8bps.dat' -ascii;
load 'ex7_8cps.dat' -ascii;
fre = ex7_8aps(:,1);
vo_25K = 1000*ex7_8aps(:,2);
vo_75K = 1000*ex7_8bps(:,2);
vo_125K = 1000*ex7_8cps(:,2);
% Determine center frequency
[vc1, k1] = max(vo_25K)
[vc2, k2] = max(vo_75K)
[vc3, k3] = max(vo_125K)
fc(1) = fre(k1);  % center frequency for circuit with
R2 = 25K
fc(2) = fre(k2);  % center frequency for circuit with
R2 = 75K
fc(3) = fre(k3);  % center frequency for circuit with
R2 = 125K
% Calculate the cut-off frequencies
vgc1 = 0.707 * vc1;   % Gain at cut-off for R2 = 25K
vgc2 = 0.707 * vc2;   % Gain at cut-off for R2 = 100K
vgc3 = 0.707 * vc3;   % Gain at cut-off for R2 = 150K
tol = 1.0e-4;          % tolerance for obtaining cut-off
l1 = k1;
while(vo_25K(l1) - vgc1) > tol
    l1 = l1 +1;
end
fhi(1) = fre(l1);    % high cut-off frequency for
R2 = 25K
l1 = k1
while(vo_25K(l1) - vgc1) > tol
    l1 = l1 - 1;
end
flow(1) = fre(l1);     % Low cut-off frequency for
R2 = 25K
l2 = k2;
while(vo_75K(l2) - vgc2) > tol
    l2 = l2 + 1;
end
fhi(2) = fre(l2);    % high cut-off frequency for
R2 = 75K
l2 = k2;
while(vo_75K(l2) - vgc2) > tol
    l2 = l2 - 1;
end
flow(2) = fre(l2);     % Low cut-off frequency for
R2 = 100K
l3 = k3;
while(vo_125K(l3) - vgc3) >tol;
    l3 = l3 + 1;
```

```
end
fhi(3) = fre(13);    % High cut-off frequency
13 = k3
while(vo_125K(13) - vgc3) > tol;
    13 = 13 - 1;
end
flow(3) = fre(13);      %low cut-off frequency
% Calculate the Quality Factor
for i = 1:3
bw(i) = fhi(i)-flow(i);
Qfactor(i) = fc(i)/bw(i);
end
% Print out results
% Center frequency, high cut-off freq, low cut-off freq
and Q factor are
fc
bw
Qfactor
% plot frequency response
plot(fre,vo_25K, fre,vo_75K, fre, vo_125K)
xlabel('Frequency, Hz')
ylabel('Gain')
title('Frequency Response of a Narrowband Filter')
```

Figure 7.22 shows the frequency response of the narrowband pass filter. The results are shown in Table 7.9.

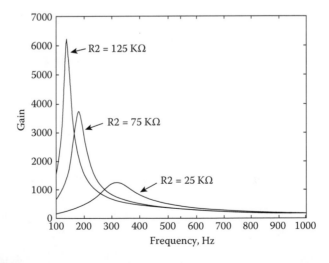

FIGURE 7.22
Narrowband filter response for various values of feedback resistance R2.

TABLE 7.9

Center Frequency, Bandwidth, and Q-Factor for a Narrowband
Pass Filter for Various Values of Feedback R2

Resistor, Ω	Center Frequency, Hz	Bandwidth, Hz	Q-Factor
25 KΩ	316.4	128.1	2.47
75 KΩ	183.0	43.3	4.23
125 KΩ	141.5	27.0	5.24

7.4.4 Band-Reject Filters

A band-reject is used to eliminate a specific band of frequencies. It is normally used in communication and biomedical instruments to eliminate
unwanted frequencies. The general form of the transfer function of the band-
reject filter is:

$$H_{BR} = \frac{k_{PB}(s^2 + \omega_C^2)}{s^2 + \left(\dfrac{\omega_C}{Q}\right)s + \omega_C^2} \tag{7.35}$$

where
k_{PB} is the passband gain; and
ω_C is the center frequency of the band-reject filter

Band-reject filters are classified as wideband reject ($Q < 10$) and narrowband reject filter ($Q > 10$). Narrowband reject filters are commonly called
notch filters. The wideband reject filter can be implemented by summing
the responses of highpass section and lowpass section through a summing
amplifier. The block diagram arrangement is shown in Figure 7.23.
   The order of the band-reject filter is dependent on the order of lowpass and
highpass sections. There are two important requirements for implementing
the wideband reject filter using the scheme shown in Figure 7.23.

1. The cut-off frequency, f_L, of the highpass filter section must be
   greater than the cut-off frequency, f_H, of the lowpass filter section;
   and the passband gains of the lowpass and highpass sections must
   be equal.

Narrowband reject filter or notch filter can be implemented by using a
Twin-T network. This circuit consists of two parallel T-shaped networks,
shown in Figure 7.24.
   Normally, R1 = R2 = R, R3 = R/2, C1 = C2 = C, and C3 = 2C. The R1-C3-R2
section is a lowpass filter with corner frequency $f_C = (4\pi RC)^{-1}$. The C1-R3-C2
is a highpass filter with corner frequency $f_C = (\pi RC)^{-1}$. The center frequency
or the notch frequency is $f_C = (2\pi RC)^{-1}$. At the latter frequency the phases of

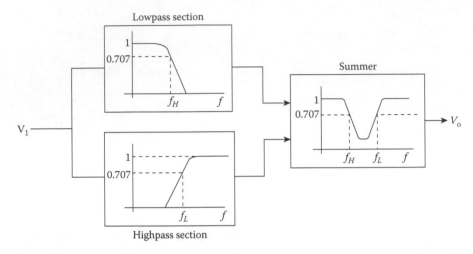

FIGURE 7.23
Block diagram of wideband reject filter.

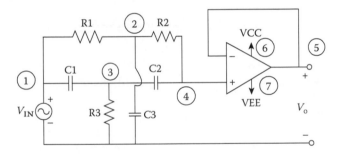

FIGURE 7.24
Narrowband-reject twin-T network.

the two filters cancel out. The following example explores the characteristics of the notch filter.

Example 7.10: Worst Case Notch Frequency

For the twin-T network shown in Figure 7.24, $R1 = R2 = 20$ KΩ, $R3 = 10$ KΩ, $C1 = C2 = 0.01$ μF, $C3 = 0.02$ μF, and the op amp is UA741. Determine (a) nominal notch frequency and (b) the worst-case notch frequency for all devices if tolerances on the resistors and capacitors are 10%. Determine the change in notch frequency due to the component tolerances.

Solution

PSPICE is used to perform Monte Carlo analysis. The worst-case notch frequency is obtained from the latter analysis. MATLAB® is used to determine the nominal and the worst-case notch frequencies. In addition, MATLAB is used to plot the frequency response of the notch filter.

PSPICE Program

```
TWIN-T BANDREJECT FILTER
.OPTIONS RELTOL = 0.10; 10% COMPONENTS
VIN  1  0   AC   1.0V
R1   1  2   RMOD   20K
R2   2  4   RMOD 20K
C3   2  0   CMOD   0.02U
C1   1  3   CMOD   0.01U
C2   3  4   CMOD   0.01U
R3   3  0   RMOD   10K
VCC  6  0   DC    15V
VEE  7  0   DC    -15V
.MODEL RMOD RES(R=1, DEV=10%);  10 % RESISTORS
.MODEL CMOD CAP(C=1, DEV=10%);  10 % CAPACITORS
X1   4  5  6   7   5  UA741; 741 OP AMP
* +INPUT; -INPUT; +VCC; -VEE; OUTPUT; CONNECTIONS FOR
UA741
.LIB NOM.LIB;
* UA741 OP AMP MODEL IN PSPICE LIBRARY FILE NOM.LIB
.AC DEC 100  10HZ   100KHZ
.WCASE AC    V(5)   MAX OUTPUT ALL; SENSITIVITY & WORST
CASE ANALYSIS
.PRINT AC    VM(5)
.PROBE V(5)
.END
```

PSPICE partial results for nominal values of the capacitors and resistors are shown in Table 7.10. The complete PSPICE output data can be found in files ex7_9aps.dat and ex7_9bps.dat for nominal and worst-case for all devices, respectively.

MATLAB is used for the data analysis.

TABLE 7.10

Frequency Response of Twin-T Notch Filter for Nominal Values of Resistors and Capacitors

Frequency, Hz	Output Voltage for 10% Device Tolerance
1.000E+01	9.987E−01
5.012E+01	9.694E−01
1.000E+02	8.905E−01
5.012E+02	2.329E−01
1.000E+03	1.145E−01
5.012E+03	8.378E−01
1.000E+04	9.523E−01
5.012E+04	9.981E−01
1.000E+05	1.000E+00

MATLAB Script

```
% Load the data
load 'ex7_9aps.dat' -ascii;
load 'ex7_9bps.dat' -ascii;
fre = ex7_9aps(:,1);
vo_nom = ex7_9aps(:,2);
vo_wc = ex7_9bps(:,2);
%
% Determination of center frequency
[vc(1), k(1)] = min(vo_nom);
[vc(2), k(2)] = min(vo_wc);
for i =1:2
fc(i) = fre(k(i));
end
% Determine difference between center frequencies
fc_dif = fc(1) - fc(2);
% Plot the frequency response
plot(fre, vo_nom, fre, vo_wc);
xlabel('Frequency, Hz')
ylabel('Gain')
title('Frequency Response of a Notch Filter')
fc
fc_dif
```

The frequency response of the notch filter is shown in Figure 7.25. The results from MATLAB are:

The center frequency of the filter with nominal component values is 794.3 Hz;

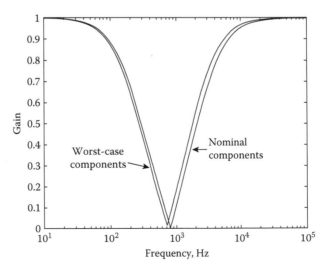

FIGURE 7.25
Frequency response of a notch filter.

The center frequency of the filter with all worst-case component values is 707.9 Hz; and
The difference between the above two center frequencies is 86.4 Hz.

Problems

7.1 For the inverting configuration shown in Figure P7.1, R1 = 1 KΩ, VCC = 15 V, and VEE = –15 V. If R2 takes values of 5 KΩ, 10 KΩ, 20 KΩ, 30 KΩ, 40 KΩ, and 50 KΩ. (a) Determine the corresponding low frequency gain and the cut-off frequency. (b) Plot a graph of cut-off frequency versus gain. (c) What is the unity gain bandwidth? Assume that the op amp is 741.

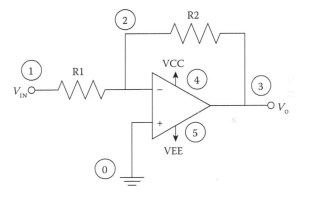

FIGURE P7.1
Inverting amplifier.

7.2 For Figure 7.5, X1 is a UA741 op amp, VCC = 15 V, VEE = –15 V, R1 = 1 KΩ, and R2 = 4 KΩ. Determine the 3-dB frequency and the unity-gain bandwidth.

7.3 A 4-pole Sallen–Key lowpass filter is shown in Figure P7.3. VCC = 15 V, VEE = –15 V, R = 3 KΩ, R1 = R3 = 10 KΩ, R2 = 150 Ω, R4 = 12 KΩ, and C = 0.01 μF. (a) Find cut-off frequency. (b) What is the gain in the passband? Assume 741 op amps.

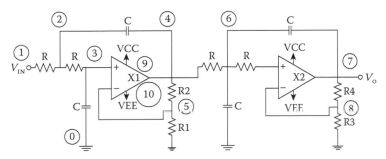

FIGURE P7.3
Sallen–Key lowpass filter.

7.4   The modified Sallen–Key circuit, shown in Figure P7.4, is similar to the Sallen–Key circuit shown in Figure 7.12, except that a voltage divider network has been added at the output. This addition results in a higher DC gain. In addition, the cut-off frequency is affected. This exercise explores the gain as a function of the cut-off frequency. Assume that R1 = R2 = 30 KΩ, R3 = 10 KΩ, R4 = 40 KΩ, C1 = C2 = 0.05 μF, R5 = 10 KΩ, VCC = 15 V, and VEE = –15 V. Op amp is 741. Determine (a) cut-off frequency. (b) Maximum gain when R6 takes the following values: (i) 10 KΩ, (ii) 8 KΩ, (iii) 6 KΩ, and (iv) 4 KΩ.

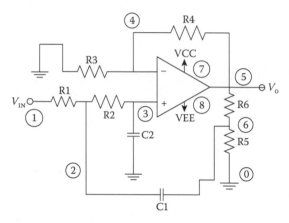

FIGURE P7.4
Modified Sallen–Key lowpass filter.

7.5   A Butterworth second-order highpass filter is shown in Figure P7.5. If C3 = 0.05 μF, R1 = R2 = R3 = 600 Ω, R4 = 400 Ω, C1 = 30 nF, and C2 = 19 nF. Determine the (a) cut-off frequency, and (b) gain at the cut-off frequency. Assume a 741 op amp.

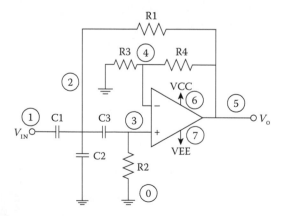

FIGURE P7.5
Butterworth highpass filter.

7.6 In Example 7.6, if R5 has the following values: 450 Ω, 500 Ω, 550 Ω, 600 Ω, and 650Ω, determine the quality factor for each value of R5. Plot quality factor as a function of R5. Assume 741 op amp.

7.7 In Figure P7.7, RG1 = RG2 = 1 KΩ, RF1 = RF2 = 3 KΩ, R1 = R2 = 45 KΩ, R3 = R4 = 100 Ω, C1 = C2 = 25 nF, and C3 = C4 = 10 nF. Assume that X1 and X2 are 741 Op amps. (a) Calculate the high frequency cutoff, (b) Determine the low frequency cutoff; (c) Calculate the bandwidth, and (d) What is the gain in the midband of the filter?

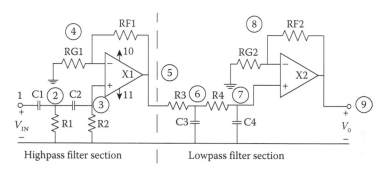

Highpass filter section | Lowpass filter section

FIGURE P7.7
Wideband pass filter implemented by cascading highpass and lowpass filters.

7.8 Figure P7.8 is a modified multiple feedback bandpass filter similar to Figure 7.21. R1 = 10 KΩ, C1 = C2 = 1 nF, and R2 = 150 KΩ. If RB is (i) 5 KΩ, (ii) 10 KΩ, and (iii) 15 KΩ, find the Q-value, center frequency and bandwidth for each value of RB. Assume that VCC = 15 V, VEE = −15 V, and the op amp is 741.

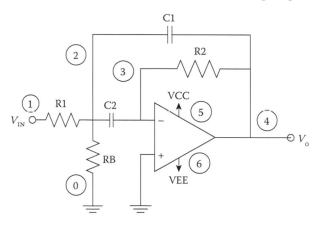

FIGURE P7.8
Modified multiple-feedback bandpass filter.

7.9 Figure P7.9 is a wideband reject filter obtained using lowpass and highpass sections. VCC = 15 V, VEE = −15 V, R3 = R5 = 1 KΩ,

R4 = R6 = 5 KΩ, R7 = R8 = R9 = 2 KΩ, R1 = 100 Ω, R2 =100 KΩ, and C1 = C2 = 0.01 μF. Determine (a) notch frequency, (b) bandwidth, and (c) quality factor. Assume that X1, X2, and X3 are UA741 op amps.

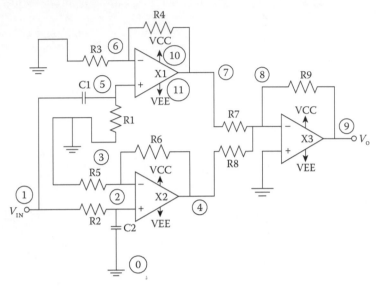

FIGURE P7.9
Wideband band-reject filter.

7.10   For the twin-T network shown in Figure 7.24, R1 = R2 = 10 KΩ, R3 = 5 KΩ, C1 = C2 = 0.01 μF, C3 = 0.02 μF, and the op amp is μA741. Determine (a) the notch frequency, (b) the bandwidth of the filter, (c) quality factor, and (d) worst-case notch frequency if the resistors and capacitors have tolerance of 5%.

7.11   In Example 7.9, the component tolerance is changed to 15%. (a) Find the notch frequency for the nominal values of the capacitors and resistors. (b) Determine the worst-case notch frequency. (c) For the nominal component values, determine the bandwidth and quality factor of the notch filter.

7.12   In Example 7.1, assume that R1 = 2 K, R2 = 10 K, and the Op amp is UA741. If $V_{CC} = |V_{EE}| = 10, 11, 12, 13, 14,$ and 15 V, plot the gain of the linear region with respect to the voltage V_{CC}.

Bibliography

1. Alexander, Charles K., and Matthew N. O. Sadiku. *Fundamentals of Electric Circuits*. 4th ed. New York: McGraw-Hill, 2009.
2. Attia, J. O. *Electronics and Circuit Analysis Using MATLAB®*. 2nd ed. Boca Raton, FL: CRC Press, 2004.

3. Boyd, Robert R. *Tolerance Analysis of Electronic Circuits Using MATLAB®*. Boca Raton, FL: CRC Press, 1999.
4. Chapman, S. J. *MATLAB® Programming for Engineers*. Tampa, FL: Thompson, 2005.
5. Davis, Timothy A., and K. Sigmor. *MATLAB® Primer*. Boca Raton, FL: Chapman & Hall/CRC, 2005.
6. Distler, R. J. "Monte Carlo Analysis of System Tolerance." *IEEE Transactions on Education* 20 (May 1997): 98–101.
7. Etter, D. M. *Engineering Problem Solving with MATLAB®*. 2nd ed. Upper Saddle River, NJ: Prentice Hall, 1997.
8. Etter, D. M., D. C. Kuncicky, and D. Hull. *Introduction to MATLAB® 6*. Upper Saddle River, NJ: Prentice Hall, 2002.
9. Hamann, J. C., J. W. Pierre, S. F. Legowski, and F. M. Long. "Using Monte Carlo Simulations to Introduce Tolerance Design to Undergraduates." *IEEE Transactions on Education* 42, no. 1 (February 1999): 1–14.
10. Gilat, Amos. *MATLAB®, An Introduction With Applications*. 2nd ed. New York: John Wiley & Sons, 2005.
11. Hahn, Brian D., and Daniel T. Valentine. *Essential MATLAB® for Engineers and Scientists*. 3rd ed. New York and London: Elsevier, 2007.
12. Herniter, Marc E. *Programming in MATLAB®*. Florence, KY: Brooks/Cole Thompson Learning, 2001.
13. Howe, Roger T., and Charles G. Sodini. *Microelectronics, An Integrated Approach*. Upper Saddle River, NJ: Prentice Hall, 1997.
14. Moore, Holly. *MATLAB® for Engineers*. Upper Saddle River, NJ: Pearson Prentice Hall, 2007.
15. Nilsson, James W., and Susan A. Riedel. *Introduction to PSPICE Manual Using ORCAD Release 9.2 to Accompany Electric Circuits*. Upper Saddle River, NJ: Pearson/Prentice Hall, 2005.
16. *OrCAD Family Release 9.2*. San Jose, CA: Cadence Design Systems, 1986–1999.
17. Rashid, Mohammad H. *Introduction to PSPICE Using OrCAD for Circuits and Electronics*. Upper Saddle River, NJ: Pearson/Prentice Hall, 2004.
18. Sedra, A. S., and K. C. Smith. *Microelectronic Circuits*. 5th ed. Oxford: Oxford University Press, 2004.
19. Spence, Robert, and Randeep S. Soin. *Tolerance Design of Electronic Circuits*. London: Imperial College Press, 1997.
20. Soda, Kenneth J. "Flattening the Learning Curve for ORCAD-CADENCE PSPICE." *Computers in Education Journal* XIV (April–June 2004): 24–36.
21. Svoboda, James A. *PSPICE for Linear Circuits*. 2nd ed. New York: John Wiley & Sons, Inc., 2007.
22. Tobin, Paul. "The Role of PSPICE in the Engineering Teaching Environment." Proceedings of International Conference on Engineering Education, Coimbra, Portugal, September 3–7, 2007.
23. Tobin, Paul. *PSPICE for Circuit Theory and Electronic Devices*. San Jose, CA: Morgan & Claypool Publishers, 2007.
24. Tront, Joseph G. *PSPICE for Basic Circuit Analysis*. New York. McGraw-Hill, 2004.
25. *Using MATLAB®, The Language of Technical Computing, Computation, Visualization, Programming, Version 6*. Natick, MA: MathWorks, Inc., 2000.
26. Yang, Won Y., and Seung C. Lee. *Circuit Systems with MATLAB® and PSPICE*. New York: John Wiley & Sons, Inc., 2007.

8

Transistor Characteristics and Circuits

In this chapter, the characteristics of bipolar junction transistors and metal oxide semiconductor field effect transistors are discussed. The sensitivity of the transistor biasing circuits is explored through PSPICE simulations and MATLAB® calculations. Frequency response of amplifiers and feedback amplifiers are also discussed.

8.1 Characteristics of Bipolar Junction Transistors

Bipolar junction transistor (BJT) consists of two pn junctions connected back-to-back. The operation of the BJT depends on the flow of both majority and minority carriers. The DC behavior of the BJT can be described by the Ebers-Moll model. The voltages of the base-emitter and base-collector junctions define the region of operation of the BJT. The regions of normal operation are forward-active, reverse-active, saturation, and cut-off. Table 8.1 shows the regions of operation based on the polarities of the base-emitter and base-collector junctions.

The forward-active region corresponds to forward biasing, the emitter-base junction and reverse biasing the base-collector junction. It is the normal operation region of BJT when employed for amplifications. In the forward-active region, the first order representation of collector current I_C and base current I_B are given as:

$$I_C = I_S \exp\left(\frac{V_{BE}}{V_T}\right)\left(1 + \frac{V_{CE}}{V_{AF}}\right) \tag{8.1}$$

and

$$I_B = \frac{I_S}{\beta_F} \exp\left(\frac{V_{BE}}{V_T}\right) \tag{8.2}$$

where
β_F is large signal forward current gain of common-emitter configuration;
V_{AF} is forward early voltage;
I_S is the BJT transport saturation current; and
V_T is the thermal voltage and is given as:

TABLE 8.1

Regions of Operation of BJT

Base-Emitter Junction	Base-Collector Junction	Region of Operation
Forward bias	Reverse-bias	Forward-active
Forward-bias	Forward-bias	Saturation
Reverse-bias	Reverse-bias	Cut-off

$$V_T = \frac{kT}{q} \qquad (8.3)$$

where
k is the Boltzmann's Constant ($k = 1.381 \times 10^{-23}$ VC/°K);
T is the absolute temperature in degrees Kelvin; and
q is the charge of an electron ($q = 1.602 \times 10^{-19}$ C).

If $V_{AF} \gg V_{CE}$, then from Equations 8.1 and 8.2, we have:

$$I_C = \beta_F I_B \qquad (8.4)$$

The saturation region corresponds to forward biasing both base-emitter and base-collector junctions. The cut-off region corresponds to reverse biasing the base-emitter and base-collector junctions. In the cut-off region, the collector and base currents are insignificant compared to those that flow when transistors are in the active-forward and saturation regions.

From Equation 8.2, the input characteristic of a forward biased base-emitter junction is similar to diode characteristics. In the following example, we explore the output characteristics of a BJT transistor.

Example 8.1: BJT Output Characteristics

Figure 8.1 shows an arrangement that can be used to find the output characteristics of transistor Q2N2222.

(i) Draw I_C versus V_{CE} for $I_B = 2$ μA, 4 μA, and 6 μA.
(ii) Calculate the output resistance (r_{CE}) as a function of V_{CE} for $I_B = 2$ μA.

Assume that R1 = R2 = R3 = 1 Ω.

Solution

PSPICE is used to obtain the data needed to obtain the current versus voltage characteristics of transistor Q2N2222. MATLAB® is used to plot the output characteristics and also to calculate the output resistance.

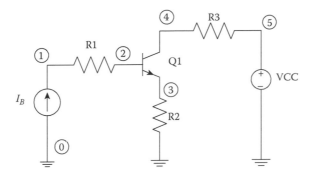

FIGURE 8.1
Circuit for obtaining BJT output characteristics.

PSPICE Program

```
BJT CHARACTERISTICS
VCC   5    0    DC    0V
R1    1    2    1
R2    3    0    1
R3    5    4    1
IB    0    1    DC    6UA
Q1    4    2    3     Q2N2222
.MODEL Q2N2222  NPN(BF = 100  IS = 3.295E-14 VA = 200);
TRANSISTOR MODEL
** ANALYSIS TO BE DONE
** VARY VCE FROM 0 TO 10V IN STEPS 0.1V
** VARY IB FROM 2 TO 6mA IN STEPS OF 2mA
.DC  VCC  0V  10V  .05V  IB  2U  6U  2U
.PRINT  DC  V(4,3)  I(R1)  I(R3)
.PROBE  V(4,3)  I(R3)
.END
```

PSPICE partial results for base current of 2 μA are shown in Table 8.2. The complete results can be found in file ex8_1aps.dat, ex8_1bps.dat, and ex8_1cps.dat for base current of 2 μA, 4 μA, and 6 μA, respectively.

MATLAB is used to plot the output characteristics.

MATLAB Script

```
% Load data
load 'ex8_1aps.dat' -ascii;
load 'ex8_1bps.dat' -ascii;
load 'ex8_1cps.dat' -ascii;
vce1 = ex8_1aps(:,2);
ic1 = ex8_1aps(:,4);
vce2 = ex8_1bps(:,2);
ic2 = ex8_1bps(:,4);
```

```
vce3 = ex8_1cps(:,2);
ic3 = ex8_1cps(:,4);
plot(vce1, ic1, vce2, ic2, vce3, ic3)
xlabel('Collector-emitter Voltage, V')
ylabel('Collector Current, A')
title('Output Characteristics')
```

The output characteristics are shown in Figure 8.2. The output resistance r_{CE} as a function of V_{CE} is obtained using MATLAB.

TABLE 8.2

Output Characteristics of Transistor Q2N2222

V_{CE}, V	I_C, A
4.996E−01	1.999E−04
1.050E+00	2.005E−04
2.000E+00	2.014E−04
3.000E+00	2.024E−04
4.000E+00	2.034E−04
5.000E+00	2.044E−04
6.000E+00	2.054E−04
7.000E+00	2.064E−04
8.000E+00	2.074E−04
9.000E+00	2.084E−04
1.000E+01	2.094E−04

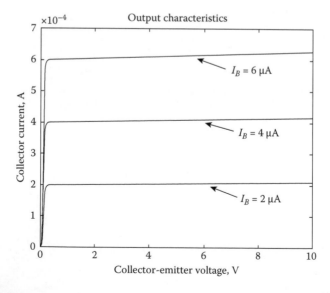

FIGURE 8.2
I_C versus V_{CE} of transistor Q2N2222.

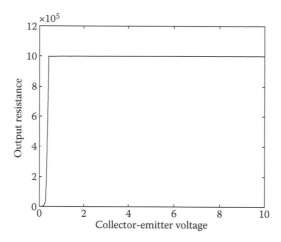

FIGURE 8.3

Output resistance as a function of collector-emitter voltage for Q2N2222 when base current is 2 μA.

MATLAB Script

```
% Load data
load 'ex8_1aps.dat' -ascii;
vce = ex8_1aps(:,2);
ic = ex8_1aps(:,4);
m = length(vce);      % size of vector vce
% calculate output resistance
for i = 2:m-1
  rce(i) = (vce(i + 1)-vce(i-1))/(ic(i + 1)- ic(i - 1));
% output resistance
end
rce(1) = rce(2);
rce(m) = rce(m-1);
plot(vce(2:m-1), rce(2:m-1))
xlabel('Collector-emitter Voltage')
ylabel('Output Resistance')
title('Output Resistance as a function of Collector-emitter
Voltage')
```

The plot of the output resistance is shown in Figure 8.3.

8.2 MOSFET Characteristics

Metal-oxide semiconductor field effect transistors (MOSFET) normally have high input resistance because of the oxide insulation between the gate and the channel. There are two types of MOSFETS: the enhancement type and the depletion type. In the enhancement type, the channel between the source and drain has to be induced by applying a voltage at the gate. In the depletion type MOSFET, the structure of the device is such that there exists a channel between the source and drain. Since the enhancement type MOSFET is widely used, the presentation of this section will be done using enhancement-type MOSFET.

The voltage needed to create the channel between the source and drain is called the threshold voltage V_T. For n-channel enhancement MOSFET, V_T is positive and for p-channel device, it is negative. MOSFETS can operate in three modes: cutoff, triode, and saturation regions. The following is a short description of the three regions of operation.

8.2.1 Cut-Off Region

For an n-channel MOSFET, if the gate-source voltage V_{GS} satisfies the condition:

$$V_{GS} < V_T \tag{8.5}$$

then the device is cut-off. This implies that the drain current is zero for all values of the drain-to-source voltage.

8.2.2 Triode Region

When $V_{GS} > V_T$ and V_{DS} is small, the MOSFET will be in the triode region. In the latter region, the device behaves as nonlinear voltage-controlled resistance. The drain current I_D is related to drain source voltage V_{DS} by:

$$I_D = k_n[2(V_{GS} - V_T)V_{DS} - V_{DS}^2](1 + \lambda V_{DS}) \tag{8.6}$$

provided

$$V_{DS} \leq V_{GS} - V_T \tag{8.7}$$

where

$$k_n = \frac{\mu_n \varepsilon \ \varepsilon_{OX}}{2t_{OX}} \frac{W}{L} = \frac{\mu_n C_{OX}}{2}\left(\frac{W}{L}\right) \tag{8.8}$$

and
   μ_n is the surface mobility of electrons;
   ε is the permitivity of free space (8.85×10^{-12} F/cm);
   ε_{OX} is the dielectric constant of SiO_2;
   t_{OX} is oxide thickness;
   L is length of the channel;
   W is width of the channel; and
   λ is channel width modulation factor.

8.2.3 Saturation Region

If $V_{GS} > V_T$, a MOSFET will operate in the saturation region provided:

$$V_{DS} \geq V_{GS} - V_T \tag{8.9}$$

In the saturation region, the current-voltage characteristics are given as:

$$I_D = k_n (V_{GS} - V_T)^2 (1 + \lambda V_{DS}) \tag{8.10}$$

The transconductance is given as:

$$g_m = \frac{\Delta I_D}{\Delta V_{GS}} \tag{8.11}$$

and the incremented drain-to-source resistance, r_{CE} is given as,

$$r_{CE} = \frac{\Delta V_{DS}}{\Delta I_{DS}} \tag{8.12}$$

In the following example, we shall obtain the I_D versus V_{GS} characteristics of a MOSFET.

Example 8.2: Current versus Voltage Characteristics of a MOSFET

Figure 8.4 shows an arrangement for obtaining the I_D versus V_{GS} characteristics of a MOSFET. (a) Draw I_D versus V_{GS} curve, (b) Obtain the transconductance versus V_{GS}. Assume that M1 is M2N4351.

Solution

PSPICE simulation is used to obtain the corresponding current versus voltage values of the MOSFET.

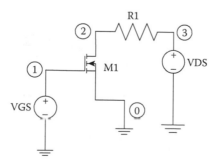

FIGURE 8.4
Circuit for obtaining MOSFET characteristics.

TABLE 8.3

I_D versus V_{GS} of MOSFET M2N4351

V_{GS}, V	I_D, A
0.000E+00	5.010E−12
5.000E−01	5.010E−12
1.000E+00	5.010E−12
1.500E+00	5.010E−12
2.000E+00	5.010E−12
2.500E+00	2.555E−05
3.000E+00	2.183E−04
3.500E+00	6.001E−04
4.000E+00	1.171E−03
4.500E+00	1.931E−03
5.000E+00	2.879E−03

PSPICE Program

```
* ID VERSUS VGS CHARACTERISTICS OF A MOSFET
VDS     3     0     DC    5V
R1      3     2     1
VGS     1     0     DC    2V;     THIS IS AN ARBITRARY
VALUE, VGS WILL BE SWEPT
M1 2 1 0 0 M2N4531;   NMOS MODEL
.MODEL M2N4531 NMOS(KP = 125U VTO = 2.24 L = 10U W = 59U
LAMBDA = 5M)

.DC     VGS    0V    5V     0.05V
** OUTPUT COMMANDS
..PRINT DC I(R1)
.PROBE V(1) I(R1)
.END
```

Partial results of the PSPICE simulation are shown in Table 8.3. The complete results can be found in file ex8_2ps.dat.

MATLAB® was used to plot I_D versus V_{GS} of the MOSFET. In addition, the transconductance versus V_{GS} is obtained using MATLAB.

MATLAB Script

```
% Load pspice data
load 'ex8_2ps.dat' -ascii;
vgs = ex8_2ps(:,1);
ids = ex8_2ps(:,2);
m = length(vgs);       % size of vector vgs
% Plot Ids versus VGS
subplot(211)
plot(vgs, ids)
ylabel('Drain Current, A')
title('Input Characteristics of a MOSFET ')
```

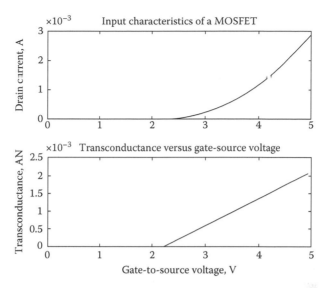

FIGURE 8.5
(a) I_D versus V_{GS}. (b) g_m versus V_{GS} of MOSFET M2N4351.

```
% Calculate transconductance
for i = 2:m - 1;
  gm(i) = (ids(i = 1) - ids(i-1))/(vgs(i + 1) - vgs(i-1)); %
transconductance
end
gm(1) = gm(2);
gm(m) = gm(m - 1);
% Plot transconductance
subplot(212);
plot(vgs(2:m - 1), gm(2:m - 1))
xlabel('Gate-to-Source Voltage, V')
ylabel('Transconductance, A/V')
title('Transconductance versus Gate-source Voltage')
```

The plots are shown in Figure 8.5.

8.3 Biasing of BJT Circuits

Biasing networks are used to establish an appropriate DC operating point for transistor in a circuit. For stable and consistent operation, the DC operating point should be held relatively constant under varying conditions. There are several biasing circuits available in the literature. Some are for biasing discrete circuits and others for integrated circuits (ICs). Figures 8.6 and 8.7 show some biasing networks for discrete circuits.

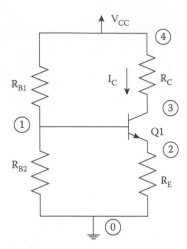

FIGURE 8.6
Biasing circuit for BJT discrete circuits with two base resistors.

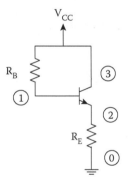

FIGURE 8.7
Biasing BJT discrete network with one base resistor.

Biasing networks for discrete circuits are not suitable for ICs because of the large number of resistors and large coupling and bypass capacitors required for biasing discrete electronic circuits. It is uneconomical to fabricate large IC resistors since they take a disproportionably large area on an IC chip. For ICs, biasing is done using mostly transistors that are connected to create constant current sources. Some biasing circuits for ICs are shown in Figures 8.8 through 8.10.

For the bias network for discrete circuits, shown in Figure 8.6, it can be shown that:

$$I_C = \frac{V_{BB} - V_{BE}}{\dfrac{R_B}{\beta_F} + \dfrac{(\beta_F + 1)}{\beta_F} R_E} \tag{8.13}$$

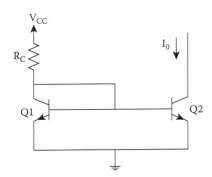

FIGURE 8.8
Simple current mirror for IC biasing.

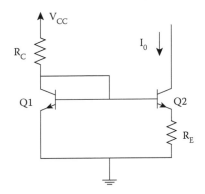

FIGURE 8.9
Widlar current source.

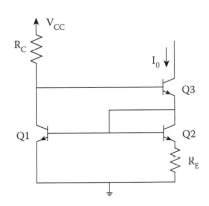

FIGURE 8.10
Wilson current source.

and

$$V_{CE} = V_{CC} - I_C \left(R_C + \frac{R_E}{\alpha_F} \right) \tag{8.14}$$

where

$$V_{BB} = \frac{V_{CC}R_{B2}}{R_{B1} + R_{B2}} \tag{8.15}$$

$$R_B = R_{B1} // R_{B2} = \frac{R_{B1}R_{B2}}{R_{B1} + R_{B2}} \tag{8.16}$$

$$\alpha_F = \frac{\beta_F}{\beta_F + 1} \tag{8.17}$$

β_F is a large signal forward current gain of common-emitter configuration.

For the simple current mirror circuit, shown in Figure 8.8, it can also be shown that:

$$I_0 = \frac{\beta_F}{\beta_F + 2} I_R \tag{8.18}$$

where

$$I_R = \frac{V_{CC} - V_{BE}}{R_C} \tag{8.19}$$

Equation 8.13 gives the parameters that influence the bias current, I_C. By using stabilized voltage supply, we can ignore changes in V_{CC} and hence V_{BB}. The changes in resistances R_B and R_E are negligible. However, there is a variation of β_F with respect to changes in I_C. In addition, there is variation of β_F of a specified transistor when selected from different lots or fabricated by different manufacturers.

DC stability of circuit when components vary can be investigated through a DC sensitivity analysis. The PSPICE command for performing sensitivity analysis is .SENS statement, which was discussed in Chapter 2. As mentioned in Chapter 2, .SENS statement allows PSPICE to compute the derivatives of pre-selected variables of the circuit to most of the components in the circuit. The following example explores the sensitivity of the bias point to components of a biasing network.

Example 8.3: Sensitivity of Collector Current to Amplifier Components

For the common-emitter biasing network, shown in Figure 8.6, $V_{CC} = 10$ V, $R_{B1} = 40$ KΩ, $R_{B2} = 10$ KΩ, $R_E = 1$ KΩ, $R_C = 6$ KΩ, and Q1 is Q2N2222. (a) Find the sensitivity of the collector current to amplifier components. (b) Two other Q2N2222 transistors are picked from a lot and they have β_F of 125 and 150. What is the change in I_C with respect to β_F. Use the following model for transistor Q2N2222:

.MODEL   Q2N2222   NPN(BF = 100 IS = 3.295E–14 VA = 200)

Solution

The bias sensitivity is obtained for QN2222 when $\beta_F = 100$.

PSPICE Program

```
* SENSITIVITY OF COLLECTOR CURRENT TO AMPLIFIER COMPONENT
VCC   4   0   DC    10V
RB1   4   1   40K
RB2   1   0   10K
RE    2   0   1K
RC    5   3   6K
VM    4   5   DC    0;   MONITOR COLLECTOR CURRENT
Q1    3   1   2     Q2N2222
.MODEL Q2N2222 NPN(BF = 100 IS = 3.295E-14 VA = 200)
* ANALYSIS TO BE DONE
.SENS I(VM)
.END
```

The following edited results are obtained from the PSPICE simulation.

```
VOLTAGE SOURCE CURRENTS
        NAME    CURRENT

        VCC     -1.460E-03
        VM       1.257E-03

DC SENSITIVITIES OF OUTPUT I(VM)
```

	ELEMENT NAME	ELEMENT VALUE	ELEMENT SENSITIVITY (AMPS/UNIT)	NORMALIZED SENSITIVITY (AMPS/PERCENT)
	RB1	4.000E+04	-3.632E-08	-1.453E-05
	RB2	1.000E+04	1.363E-07	1.363E-05
	RE	1.000E+03	-1.139E-06	-1.139E-05
	RC	6.000E+03	-7.796E-10	-4.678E-08
	VCC	1.000E+01	1.800E-04	1.800E-05
	VM	0.000E+00	-6.202E-07	0.000E+00
Q1				
	RB	0.000E+00	0.000E+00	0.000E+00
	RC	0.000E+00	0.000E+00	0.000E+00
	RE	0.000E+00	0.000E+00	0.000E+00
	BF	1.000E+02	1.012E-06	1.012E-06
	ISE	0.000E+00	0.000E+00	0.000E+00
	BR	1.000E+00	-2.692E-13	-2.692E-15
	ISC	0.000E+00	0.000E+00	0.000E+00
	IS	3.295E-14	7.044E+08	2.321E-07
	NE	1.500E+00	0.000E+00	0.000E+00
	NC	2.000E+00	0.000E+00	0.000E+00
	IKF	0.000E+00	0.000E+00	0.000E+00
	IKR	0.000E+00	0.000E+00	0.000E+00
	VAF	2.000E+02	-1.730E-09	-3.460E-09
	VAR	0.000E+00	0.000E+00	0.000E+00

TABLE 8.4

I_C Versus β_F

β_F	I_C
100	1.257 mA
125	1.278 mA
150	1.292 mA

The above simulation is performed with two values of β_F, 125 and 150. Table 8.4 shows I_C versus β_F. It can be seen from Table 8.4 that as β_F increases, I_C increases.

8.3.1 Temperature Effects

Temperature changes cause two transistor parameters to change. These are (1) base-emitter voltage (V_{BE}) and (2) collector leakage current between the base and collector (I_{CBO}). For silicon transistors, the voltage V_{BE} varies almost linearly with temperature as:

$$\Delta V_{BE} \cong -2(T_2 - T_1)mV \tag{8.20}$$

where
T_1 and T_2 are temperatures in degrees Celsius.

The collector-to-base leakage current, I_{CBO}, approximately doubles every 10°C temperature rise. From Equations 8.13, 8.18, and 8.19, both I_C and I_O are dependent on V_{BE}. Thus, the bias currents are temperature dependent. The following example explores the sensitivity of the collector current to temperature variation.

Example 8.4: Sensitivity to Temperature of Common-Collector Amplifier

In Figure 8.7, RB = 40 KΩ, RE = 2 KΩ, Q1 is Q2N3904, and VCC = 10 V. Assume a linear dependence of resistance RB and RE on temperature and TC1 = 1000 ppm/°C. Determine the emitter current as a function of temperature (0°C to 100°C).

Solution

PSPICE .DC TEMP command will be used to sweep temperature from 0°C to 100°C in steps of 10°C. The changes in resistance due to temperature are calculated by SPICE using the equation:

$$R[T_2] = R(T_1)\left[1 + TC1(T_2 - T_1) + TC2(T_2 - T_1)^2\right] \tag{8.21}$$

where
$T_1 = 27°C$;
T_2 is the required temperature; and
TC1 and TC2 are included in the model statement of resistors.

PSPICE Program

```
EMITTER CURRENT DEPENDENCE ON TEMPERATURE
VCC   3   0   DC      10V
RB    3   1   RMOD3   40K;  RB IS MODELED
RE    2   0   RMOD3   2K;   RE IS MODELED
.MODEL RMOD3 RES(R=1 TC1=1000U TC2=0);    TEMP MODEL OF
RESISTORS
Q1    3   1   2       Q2N3904;      TRANSISTOR CONNECTIONS
.MODEL Q2N3904 NPN(IS=1.05E-15 ISE=4.12N NE=4 ISC=4.12N
NC=4 BF=220
+ IKF=2E-1 VAF=80 CJC=4.32P CJE=5.27P RB=5 RE=0.5 RC=1
TF=0.617N
+ TR=200N KF=1E-15 AF=1)
* ANALYSIS TO BE DONE
.DC TEMP 0 100 5;     VARY TEMP FROM 0 TO 100 IN STEPS OF 5
.PRINT DC I(RE)
.END
```

Table 8.5 shows PSPICE partial results. The complete data can be found in file ex8_4ps.dat. MATLAB® is used to plot the PSPICE results.

MATLAB Script

```
% Load data
load 'ex8_4ps.dat' -ascii;
temp = ex8_4ps(:,1);
ie = ex8_4ps(:,2);
% plot ie versus temp
plot(temp, ie, temp, ie, 'ob')
xlabel('Temperature, °C')
ylabel('Emitter Current, A')
title('Variation of Emitter Current with Temperature')
```

The plot is shown in Figure 8.11.

TABLE 8.5

Temperature versus Emitter Current

Temperature, °C	Emitter Current, A
0.000E+00	4.230E−03
1.000E+01	4.193E−03
2.000E+01	4.156E−03
3.000E+01	4.121E−03
4.000E+01	4.086E−03
5.000E+01	4.053E−03
6.000E+01	4.019E−03
7.000E+01	3.987E−03
8.000E+01	3.955E−03
9.000E+01	3.924E−03
1.000E+02	3.894E−03

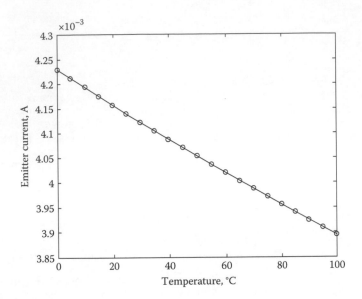

FIGURE 8.11
Emitter current as a function of temperature.

8.4 MOSFET Bias Circuit

There are several circuits that can be used to bias MOSFETS at a stable operating point such that the bias point does not change significantly with changes in the MOSFETS parameters. Some biasing circuits for MOSFETS discrete circuits are shown in Figures 8.12 through 8.14.

In Figure 8.12, it can be shown:

$$V_{GS} = \frac{R_{G1}}{R_{G1} + R_{G2}} V_{DD} - I_D R_S \tag{8.22}$$

and

$$V_{DS} = V_{DD} - I_D (R_D + R_S) \tag{8.23}$$

For integrated circuit MOSFET, constant current sources are used for biasing. A basic MOSFET current source is shown in Figure 8.15. For Figure 8.15, it can be shown that I_0 is related to I_{REF} through the expression:

$$I_0 = \frac{(W/L)_2}{(W/L)_1} I_{REF} \tag{8.24}$$

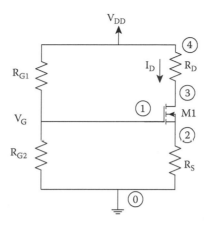

FIGURE 8.12
Biasing circuit for MOSFET using fixed gate voltage and self-bias resistors, R_S.

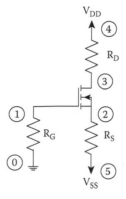

FIGURE 8.13
MOSFET biasing circuit using two power supplies.

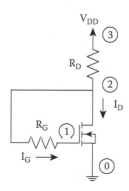

FIGURE 8.14
Biasing MOSFET circuit with resistance feedback using resistor, R_G.

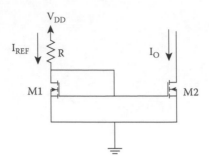

FIGURE 8.15
Basic MOSFET current source.

and

$$I_{REF} = \frac{V_{DD} - V_{GS}}{R} \tag{8.25}$$

where
$(W/L)_2$ is ratio of width to length of transistor Q2; and
$(W/L)_1$ is ratio of width to length of transistor Q1.

Thus, the current I_0 depends on the transistor sizing. The follow example explores the changes in the drain current as the source resistor is changed.

Example 8.5: Effect of Source Resistance on MOSFET Operating Point

For the MOSFET biasing circuit, shown in Figure 8.12, $V_{DD} = 10$ V, $R_{G1} = R_{G2} = 9$ MΩ, and $R_D = 8$ KΩ. Find drain current when R_S is varied from 5 KΩ to 10 KΩ in steps of 1 KΩ. Assume that M1 is M2N4351.

Solution
PSPICE is used to obtain the drain current as source current is varied. The .STEP command will be used to vary the source resistance.

PSPICE Program

```
MOSFET BIAS CIRCUIT
VDD   4   0   DC        10V;   SOURCE VOLTAGE
RG1   4   1   9.0E6;
RG2   1   0   9.0E6;
RD    4   3   8.0E3
M1    3   1   2    2    M2N4351; NMOS MODEL
RS    2   0   RMOD3   1
.MODEL RMOD3 RES(R = 1)
```

```
.STEP   RES     RMOD3 (R) 5000 10000 1000;   VARY RS FROM 5K
TO 10K
.MODEL M2N4351 NMOS (KP = 125U VTO = 2.24 L = 10U W = 59U
LAMBDA = 5M)
.DC     VDD 10 10 1
.PRINT DC      I(RD)   V(3,2)
.END
```

Table 8.6 shows the drain current for various values of source resistance. The data can also be found in file ex8_5ps.dat.

MATLAB® is used to plot I_D versus R_S.

MATLAB Script

```
% Load data
load 'ex8_5ps.dat' -ascii;
rs = ex8_5ps(:,1);
id = ex8_5ps(:,2);
plot(rs,id)
xlabel('Source Resistance')
ylabel('Drain Current')
title('Source Resistance versus Drain Current')
```

Figure 8.16 shows the drain current as a function of source resistance. Figure 8.16 shows that as the source resistance increases, the drain current reduces. The following example explores the worst-case bias point with device tolerance.

Example 8.6: Worst-Case Drain Current of a MOSFET Biasing Circuit

For the MOSFET biasing circuit shown in Figure 8.13, $V_{DD} = 5$ V, $V_{SS} = -5$ V, $R_G = 10$ MΩ, and $R_S = R_D = 4$ KΩ. Determine the worst-case bias current when the tolerances of the resistors are 1%, 2%, 5%, 10%, and 15%, respectively. Assume that M1 is M2N4351.

Solution

PSPICE .WCASE command is used to perform the worst-case analysis.

TABLE 8.6

Drain Current versus Source Resistance

Source Resistance R_S	Drain Current I_D
5000	3.576E−04
6000	3.094E−04
7000	2.731E−04
8000	2.447E−04
9000	2.218E−04
10K	2.030E−04

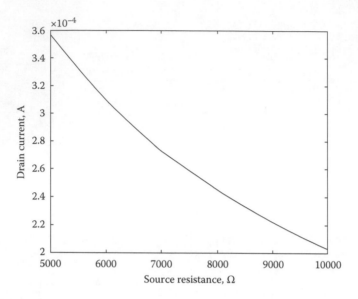

FIGURE 8.16
Drain current versus source resistance.

PSPICE Program

```
* MOSFET BIASING CIRCUIT
.OPTIONS RELTOL = 0.01;   1% COMPONENT TOLERANCE,
* CHANGED FOR DIFFERENT TOLERANCE VALUES
VSS   5   0   DC      -5V
VDD   4   0   DC      5V
RG    1   0   RMOD    10.0E6
RS    2   5   RMOD    4.0E3
RD    4   3   RMOD    4.0E3
.MODEL RMOD RES(R = 1 DEV = 1%); 1% RESISTOR TOLERANCE.
* CHANGE FOR DIFFERENT TOLERANCE VALUES
M1 3 1 2 2 M2N4351
.MODEL M2N4351 NMOS (KP = 125U VTO = 2.24 L = 10U W = 59U
LAMBDA = 5M)
.DC      VDD     5       5       1
.WCASE DC I(RD) MAX OUTPUT ALL;     WORST CASE ANALYSIS
.END
```

To obtain results for the 2% component tolerance, the relevant two statements in the above PSPICE Program are changed to:

.OPTION   RELTOL = 0.02;   2% component tolerance

.MODEL   RMOD   RES(R = 1 DEV = 2%).

TABLE 8.7

Device Tolerance versus Worst-Case Drain Current

Device Tolerance in Percent	Worst-Case (All Devices) Drain Current, A
0	425.71E−06
1	428.98E−06
2	432.31E−06
5	442.62E−06
10	460.73E−06
15	480.81E−06

The above statements are appropriately changed for 5%, 10%, and 15% component tolerance.

Table 8.7 shows the worst-case (all devices) drain current versus component tolerance in percent. The data can be found in file ex8_6ps.dat. MATLAB® is used to plot the data in Table 8.7.

MATLAB Script

```
% Load data
load 'ex8_6ps.dat' -ascii;
tol = ex8_6ps(:,1);
id_wc = ex8_6ps(:,2);
% plot data
plot(tol, id_wc, tol,id_wc,'ob')
xlabel('Device Tolerance, %')
ylabel('Worst-case Drain Current, A')
title('Worst-case Drain Current as a Function of Device
Tolerance')
```

The plot is shown in Figure 8.17. Figure 8.17 shows that the worst-case drain current increases as the device tolerance increases.

8.5 Frequency Response of Transistor Amplifiers

Amplifiers are normally used for voltage amplification, current amplification, impedance matching, or isolation between stages. Transistor amplifiers can be built using BJT and/or field effect transistors. Amplifiers built using BJT can be common-emitter, common collector (emitter follower), or common-base amplifier. Common-emitter amplifiers have relatively high voltage gain. A common-collector amplifier has relatively high input resistance, low output resistance with voltage gain that is almost equal to unity. Common-drain

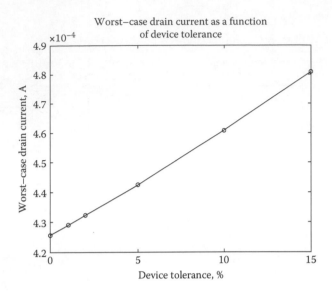

FIGURE 8.17
Device tolerance versus worst case drain current.

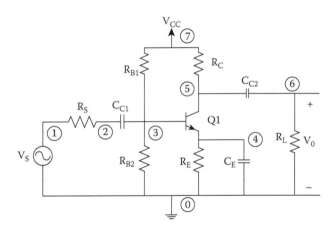

FIGURE 8.18
Common-emitter amplifier.

amplifiers have relatively low input resistance. FET amplifiers can be common-source, common-drain, or common-drain common-gate amplifier.

A common-emitter amplifier is shown in Figure 8.18. The amplifier is capable of generating a relatively high current and voltage gains. The input resistance is medium and is essentially independent of the load resistance R_L.

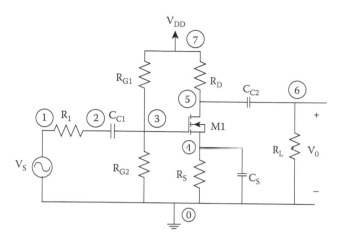

FIGURE 8.19
Common-source amplifier.

The coupling capacitor, C_{C1}, couples the voltage source, V_S, to the bias network. Coupling capacitor, C_{C2}, connects the collector resistance, R_C, to the load R_L. The bypass capacitance, C_E, is used to increase the midband gain, since it effectively short circuits the emitter resistance R_E at midband frequencies. The resistance R_E is needed for bias stability. The external capacitors C_{C1}, C_{C2}, and C_E, will influence the low frequency response of the common emitter amplifier. The transistor internal capacitances of the transistor will control the high frequency cut-off.

The common-source amplifier, shown in Figure 8.19, has characteristics similar to those of the common-emitter amplifier. However, the common-source amplifier has higher input resistance than that of the common-emitter amplifier.

The external capacitors C_{C1}, C_{C2}, and C_S, will influence the low frequency response. The internal capacitances of the FET will affect the high frequency response. In the following example, the gain and bandwidth, as a function of power supply voltage, of the common-source amplifiers are explored.

Example 8.7: Common-Source Amplifier Characteristics

In the common-source amplifier, shown in Figure 8.19, $C_{C1} = C_{C2} = 0.05$ μF, $C_S = 1000$ μF, $R_D = 6$ KΩ, $R_L = 10$ KΩ, $R_S = 2$ KΩ, $R_1 = 50$ Ω, $R_{G1} = 10$ MΩ, and $R_{G2} = 10$ MΩ. The MOSFET is IRF150. Determine the midband gain, low cutoff frequency, and bandwidth as the power supply V_{DD} varies from 6 V to 10 V.

Solution

PSPICE is used to obtain the frequency response as the power supply is varied. The PSPICE command .STEP will be used to vary the power supply voltage.

PSPICE Program

```
* COMMON-SOURCE AMPLIFIER
VDD      7    0    DC    8V
.STEP    VDD  6    10    1
R1       1    2    50
CC1      2    3    0.05UF
RG2      3    0    10MEG
RG1      7    3    10MEG
RS       4    0    2K
CS       4    0    1000UF
RD       7    5    6K
VS       1    0    AC    1MV
CC2      5    6    0.05UF
RL       6    0    10K
M1       5    3    4     4      IRF150
.LIB NOM.LIB;
* IRF 150 MODEL IN PSPICE LIBRARY FILE NOM.LIB
* AC ANALYSIS
.AC      DEC  20   10    10MEGHZ
.PRINT AC     VM(6)
.PROBE V(6)
.END
```

PSPICE results for supply voltages of 6 V, 7 V, 8 V, 9 V, and 10 V are stored in files ex8_7aps.dat, ex8_7bps.dat, ex8_7cps.dat, ex8_7dps.dat, and ex8_7eps. dat, respectively. MATLAB® is used to analyze the PSPICE results and to plot the frequency response.

MATLAB Script

```
% Load data
load 'ex8_7aps.dat' -ascii;
load 'ex8_7bps.dat' -ascii;
load 'ex8_7cps.dat' -ascii;
load 'ex8_7dps.dat' -ascii;
load 'ex8_7eps.dat' -ascii;
%
fre = ex8_7aps(:,1);
vo_6V = 1000*ex8_7aps(:,2);
vo_7V = 1000*ex8_7bps(:,2);
vo_8V = 1000*ex8_7cps(:,2);
vo_9V = 1000*ex8_7dps(:,2);
vo_10V = 1000*ex8_7eps(:,2);
% Determine center frequency
[vc1, k1] = max(vo_6V)
[vc2, k2] = max(vo_7V)
[vc3, k3] = max(vo_8V)
[vc4, k4] = max(vo_9V)
[vc5, k5] = max(vo_10V)
%
```

```
fc(1) = fre(k1);      % center frequency for VDD = 6V
fc(2) = fre(k2);      % center frequency for VDD = 7V
fc(3) = fre(k3);      % center frequency for VDD = 8V
fc(4) = fre(k4);      % center frequency for VDD = 9V
fc(5) = fre(k5);      % center frequency for VDD = 10V
% Calculate the cut-off frequencies
vgc1 = 0.707*vc1;     % Gain at cut-off for VDD = 6V
vgc2 = 0.707*vc2;     % Gain at cut-off for VDD = 7V
vgc3 = 0.707*vc3;     % Gain at cut-off for VDD = 8V
vgc4 = 0.707*vc4;     % Gain at cut-off for VDD = 9V
vgc5 = 0.707*vc5;     % Gain at cut-off for VDD = 10V
%
tol=1.0e-5;   % tolerance for obtaining cut-off
%
l1 = k1;
while(vo_6V(l1) - vgc1) > tol
      l1 = l1 + 1;
end
fhi(1) = fre(l1);     % high cut-off frequency for VDD = 6V
l1 = k1
while(vo_6V(l1) - vgc1) > tol
      l1 = l1 - 1;
end
flow(1) = fre(l1);    % Low cut-off frequency for VDD = 6V
%
l2 = k2;
while(vo_7V(l2) - vgc2) > tol
      l2 = l2 + 1;
end
fhi(2) = fre(l2);     % high cut-off frequency for VDD = 7V
l2 = k2;
while(vo_7V(l2) - vgc2) > tol
      l2 = l2 - 1;
end
flow(2) = fre(l2);    % Low cut-off frequency for VDD = 7V
%
l3 = k3;
while(vo_8V(l3) - vgc3) > tol;
      l3 = l3 + 1;
end
fhi(3) = fre(l3);     % High cut-off frequency for VDD = 8V
l3 = k3
while(vo_8V(l3) - vgc3) > tol;
      l3 = l3 - 1;
end
flow(3) = fre(l3);    %low cut-off frequency for VDD = 8V
%
l4 = k4;
while(vo_9V(l4) - vgc4) > tol;
      l4 = l4 + 1;
end
```

```
fhi(4) = fre(14);      % High cut-off frequency for VDD = 9V
14 = k4
while(vo_9V(14) - vgc4) > tol;
      14 = 14 - 1;
end
flow(4) = fre(14);     %low cut-off frequency for VDD = 9V
%
15 = k5;
while(vo_10V(15) - vgc5) > tol;
      15 = 15 + 1;
end
fhi(5) = fre(15);      % High cut-off frequency for VDD = 10V
15 = k5
while(vo_10V(15) - vgc5) > tol;
      15 = 15 - 1;
end
flow(5) = fre(15);     %low cut-off frequency for VDD = 10V
%
% Calculate the Quality Factor
for i = 1:5
bw(i) = fhi(i)-flow(i);
Qfactor(i) = fc(i)/bw(i);
end
%midband gain
gain_mb = [vc1 vc2 vc3 vc4 vc5];
% Print out results
% Gain Center frequency, high cut-off freq, low cut-off
freq and Q factor are
gain_mb
flow
bw
Qfactor

% plot frequency response
plot(fre,vo_6V, fre,vo_7V, fre, vo_8V,fre,vo_9V,fre,vo_10V)
xlabel('Frequency, Hz')
ylabel('Gain')
title('Frequency Response of a Common-source Amplifier')
```

The gain, low cut-off frequency, and bandwidth are shown in Figure 8.20. Table 8.8 shows the results obtained from MATLAB. Table 8.8 shows that as the supply voltage increases, the midband gain and the low cut-off frequency increase.

Example 8.8: Input Resistance of Emitter Follower

Figure 8.21 shows an emitter follower circuit. RS = 100 Ω, RB1 = 80 KΩ, VCC = 15 V, and C1 = 5 μF. If RE changes from 500 to 2000 Ω, determine the input resistance as a function of emitter resistance RE. In addition, determine the changes in the collector-emitter voltage as RE varies. Assume that Q1 is Q2N2222.

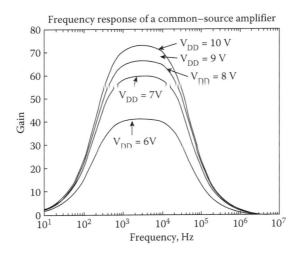

FIGURE 8.20
Frequency response of a common-source amplifier as different supply voltages.

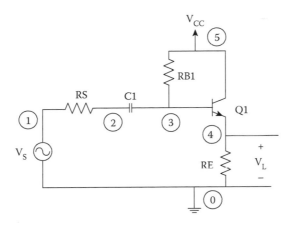

FIGURE 8.21
Emitter follower circuit.

TABLE 8.8

Gain, Low Cut-Off Frequency and Bandwidth as a Function of Supply Voltage

Supply Voltage, V	Midband Gain	Low Cut-Off Frequency, Hz	Bandwidth, Hz
6	41.29	223.9	3.5256E+04
7	59.84	251.2	3.9559E+04
8	66.40	251.2	3.9559E+04
9	70.21	251.2	3.9559E+04
10	72.79	281.8	3.5198E+04

Solution

The .STEP statement is used to vary the emitter resistance RE in the PSPICE program.

PSPICE Program

```
* INPUT RESISTANCE OF AN EMITTER FOLLOWER
VS       1       0       AC      10E-3
VCC      5       0       DC      15V
RS       1       2       100
C1       2       3       5UF
RB       5       3       80K
RE       4       0       RMOD4   1
.MODEL RMOD4 RES(R = 1)
.STEP RES RMOD4(R) 500        2000    150; VARY RE FROM 500 TO
2000
Q1       5       3       4       Q2N2222
.MODEL Q2N2222 NPN(BF = 100 IS = 3.295E-14 VA = 200);
TRANSISTOR MODEL
.DC      VCC     15      15      1
.AC      LIN     1       1000    1000
.PRINT DC        I(RE)   V(5,4)
.PRINT AC        V(1)    I(RS)
.END
```

The results obtained from PSPICE are shown in Table 8.9. The complete results are also available in file ex8_8ps.dat. MATLAB® is used obtain the input imped-ance and also to plot the results.

MATLAB Script

```
% Load data
load 'ex8_8ps.dat' -ascii;
re = ex8_8ps(:,1);
ie_dc = ex8_8ps(:,2);
vce_dc = ex8_8ps(:,3);
ib_ac = ex8_8ps(:,4);
vs_ac = 10.0e-03; % input signal is 10 mA
% Calculate input resistance
m = length(re);
for i = 1:m
 rin(i) = vs_ac/ib_ac(i);
end
subplot(211), plot(re, rin, re, rin,'ob')
ylabel('Input Resistance, Ohms')
title('(a) Input Resistance')
subplot(212), plot(re, vce_dc, re, vce_dc, 'ob')
ylabel('Collector-emitter Voltage, V')
title('(b) DC Collector-Emitter Voltage')
xlabel('Emitter Resistance, Ohms')
```

The input resistance and collector-emitter voltage are shown in Figure 8.22.

TABLE 8.9

(a) DC Emitter Current, (b) DC Collector-Emitter Voltage, and (c) Small-Signal Input Current versus Emitter Resistance

Emitter Resistance, Ω	DC Emitter Current, A	DC Collector-to-Emitter Voltage, V	Small-Signal Input Current, A
500	1.136E−02	9.318	3.181E−07
650	1.013E−02	8.413	2.750E−07
800	9.147E−03	7.683	2.478E−07
950	8.338E−03	7.079	2.290E−07
1100	7.662E−03	6.572	2.153E−07
1250	7.088E−03	6.140	2.048E−07
1400	6.595E−03	5.767	1.965E−07
1550	6.166E−03	5.442	1.898E−07
1700	5.791E−03	5.156	1.842E−07
1850	5.458E−03	4.902	1.795E−07
2000	5.162E−03	4.675	1.755E−07

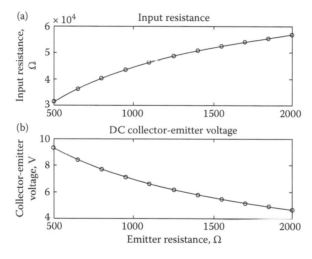

FIGURE 8.22

(a) Input resistance versus emitter resistance. (b) Collector-emitter voltage versus emitter resistance.

8.6 Schematic Capture of Transistor Circuits

The ORCAD CAPTURE can be used to draw and simulate transistor circuits. Start the ORCAD schematic using the steps outlined in Box 1.1. Draw the transistor circuit using the steps outlined in Box 1.2. In the latter box, choose the transistor by selecting the transistor part. In the student version of the ORCAD CAPTURE, select the transistor from the BREAKOUT or EVAL

> **BOX 8.1   SEQUENCE OF STEPS FOR SIMULATING TRANSISTOR CIRCUITS**
>
> - Use the steps in Box 1.1 to start ORCAD schematic.
> - Use the steps in Box 1.2 to draw the circuit using ORCAD schematic.
> - In the steps in Box 1.2, you can select transistor part. In the Student version of the ORCAD Schematic package, you can select the transistor from the BREAKOUT or EVAL library.
> - The transistor will have model parameters. To set the parameters of the device, select the device and go to Edit > PSPICE Model. Click it to open the PSPICE Model Editor and insert the model for the device. Model of various devices are available on the Internet from device manufacturers' websites.
> - Use Boxes 1.3, 1.4, 1.5, and 1.6 to perform DC, DC Sweep, Transient, and AC Analysis, respectively.

library. DC, DC Sweep, Transient analysis, and AC analysis can be performed by using the steps outlined in Boxes 1.3, 1.4, 1.5, and 1.6, respectively. Box 8.1 shows the steps needed to perform the analysis of diode circuits.

Example 8.9: Frequency Response of a Common-Emitter Amplifier

In the common-emitter amplifier shown in Figure 8.18, $C_{C1} = C_{C2} = 2$ µF, $C_E = 100$ µF, $R_{B1} = 100$ KΩ, $R_{B2} = 100$ KΩ, $R_S = 50$ Ω, $R_L = R_C = 8$ KΩ, $R_E = 1.5$ KΩ, and VCC = 10 V. Transistor Q1 is Q2N2222. Plot the output voltage with respect to frequency.

Solution

Figure 8.18 was drawn using the schematic capture. Figure 8.23 shows the schematic of the circuit. AC analysis was performed. Figure 8.24 shows the output voltage.

8.7 Feedback Amplifiers

The general structure of a feedback amplifier is shown in Figure 8.25. *A* is the open-loop gain of the amplifier without feedback. Input X_I and output X_0 are related by:

$$X_0 = AX_i \qquad (8.26)$$

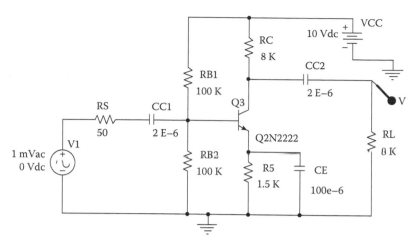

FIGURE 8.23
Common-emitter amplifier.

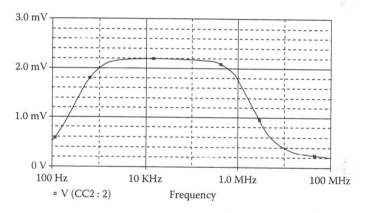

FIGURE 8.24
Output voltage of the common-emitter amplifier.

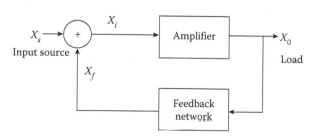

FIGURE 8.25
General structure of feedback amplifier.

The output quantity X_0 is fed back to the input through the feedback network, which provides a sample signal X_f. The latter is related to the output X_0 by the expression:

$$X_f = \beta X_0 \tag{8.27}$$

The feedback signal X_f is subtracted from the source (for negative feedback amplifier) to produce the input signal:

$$X_i = X_S - X_f \tag{8.28}$$

For positive feedback, the latter equation, becomes:

$$X_i = X_S + X_f \tag{8.29}$$

Combining Equations 8.26 to 8.28, we get:

$$A_f = \frac{x_O}{x_S} = \frac{A}{1 + \beta A} \tag{8.30}$$

where
β_A is the loop gain; and
$(1 + \beta A)$ is the amount of feedback.

It can be shown that amplifiers with negative feedback will result in (i) gain insensitivity to component variations, (ii) increased bandwidth, and (iii) reduced nonlinear distortion. The feedback amplifier topologies are shown in Figure 8.26.

Depending on the type of feedback configuration, input and output resistance can be shown to increase or decrease by the feedback factor β. Table 8.10 shows the feedback relationships.The following two examples will explore the characteristics of feedback amplifiers.

Example 8.10: Two-Stage Amplifier with Feedback Resistance

The circuit shown in Figure 8.27 is an amplifier with shunt-shunt feedback. RB1 = RB2 = 50 KΩ, RS = 100 Ω, RC1 = 5 KΩ, RE1 = 2.5 KΩ, RC2 = 10 KΩ, RE2 = 2 KΩ, C1 = 20μF, CE2 = 100 μF, and VCC = 15 V. If Vs = 1 mV, find V_0 as R_F changes from 1 KΩ to 8 KΩ. Plot V_0 versus R_F. Assume that both transistors Q1 and Q2 are Q2N2222. The source voltage is a 2 KHz sine wave with 1 mV peak voltage.

Solution

PSPICE is used to obtain the output voltage with respect to R_F. MATLAB® is used to obtain the relationship between output voltage and R_F.

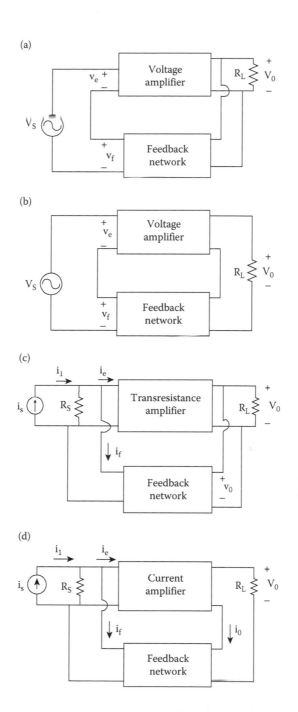

FIGURE 8.26
Feedback configurations. (a) Series-shunt feedback, (b) Series-series feedback, (c) Shunt-resistor feedback, and (d) Shunt-series feedback.

TABLE 8.10

Feedback Relationships

Amplifier Configuration	Gain	Input Resistance	Output Resistance
Without resistance	A	R_i	R_O
Series-shunt amplifier	$A_f = \dfrac{A}{1+\beta A}$	$R_{if} = R_i(1+\beta A)$	$R_{of} = \dfrac{R_O}{1+\beta A}$
Series-series amplifier	$A_f = \dfrac{A}{1+\beta A}$	$R_{if} = R_i(1+\beta A)$	$R_{of} = R_O(1+\beta A)$
Shunt-shunt amplifier	$A_f = \dfrac{A}{1+\beta A}$	$R_{if} = \dfrac{R_i}{1+\beta A}$	$R_{of} = \dfrac{R_O}{1+\beta A}$
Shunt-series amplifier	$A_f = \dfrac{A}{1+\beta A}$	$R_{if} = \dfrac{R_i}{1+\beta A}$	$R_{of} = R_O(1+\beta A)$

PSPICE Program

```
AMPLIFIER WITH FEEDBACK
VS    1    0    AC    1MV    0
RS    1    2    100
C1    2    3    20E-6
RB1   3    0    50E3
RB2   6    3    50E3
RE1   4    0    2.5E3
RC1   6    5    5.0E3
Q1    5    3    4      Q2N2222
.MODEL Q2N2222 NPN(BF = 100 IS = 3.295E-14 VA = 200);
TRANSISTORS MODEL
VCC   6    0    DC    15V
RE2   7    0    2E3
CE2   7    0    100E-6
RC2   6    8    10.0E3
Q2    8    4    7      Q2N2222
.AC   LIN  1    2000   2000
RF    8    3    RMODF 1
.MODEL RMODF RES(R = 1)
.STEP   RES    RMODF(R) 1.0E3 8.0E3 1.0E3
.PRINT AC VM(8,0)
.END
```

Table 8.11 shows the results obtained from PSPICE. The results are also available in file ex8_9ps.dat. MATLAB is used to plot the results.

MATLAB Script

```
%load data
[rf, gain] = textread('ex8_9ps.dat', '%d %f');
plot(rf, gain, rf, gain,'ob')
title('Gain versus Feedback Resistance')
xlabel('Feedback Resistance, Ohms')
ylabel('Gain')
```

The feedback resistance versus gain is shown in Figure 8.28.

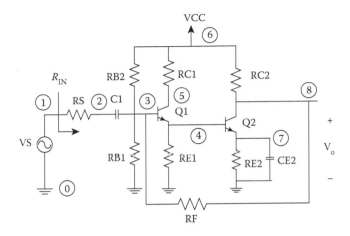

FIGURE 8.27
Amplifier with shunt-shunt feedback.

TABLE 8.11

Gain versus Feedback Resistance

RF, Ω	Gain
1000	7.932
2000	15.78
3000	23.38
4000	30.75
5000	37.91
6000	44.86
7000	51.62
8000	57.01

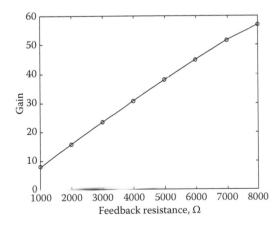

FIGURE 8.28
Feedback resistance versus gain of a two-stage amplifier.

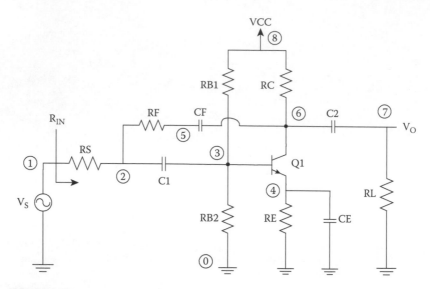

FIGURE 8.29
Common-emitter amplifier with shunt-shunt feedback.

Example 8.11: Common-Emitter Amplifier with Feedback Resistances

In Figure 8.29, RS = 150 Ω, RB2 = 20 KΩ, RB1 = 90 KΩ, RE = 2 KΩ, RC = 5 KΩ, RL = 10 KΩ, C1 = 2 μF, CE = 50 μF, C2 = 2 μF, CF = 5 μF, and VCC = 15 V. Find input resistance, R_{IN} and the voltage gain as the feedback resistance changes from 1 to 10 KΩ. The source voltage is a 1 KHz sine wave with 1 mV peak voltage. Assume that Q1 is Q2N2222.

Solution

PSPICE is used to perform circuit simulation. The PSPICE command .STEP is used to vary the feedback resistance.

PSPICE Program

```
COMMON  EMITTER  AMPLIFIER
VS      1       0       DC      0       AC      1E-3    0
RS      1       2       150
C1      2       3       2E-6
RB2     3       0       20E3
RB1     8       3       90E3
CF      5       6       5E-6
Q1      6       3       4       Q2N2222
```

```
.MODEL Q2N2222 NPN(BF = 100 IS = 3.295E-14 VA = 200);
TRANSISTORS MODEL
RE      4       0       2.0E3
CE      4       0       50E-6
RC      8       6       5.0E3
C2      6       7       2.0E-6
RL      7       0       10.0E3
VCC     8       0       DC      15V
RF      2       5       RMODF   1
.MODEL RMODF RES(R = 1)
.STEP RES RMODF (R) 1.0E3 10E3 1.0E3;
.AC     LIN     1       1000    1000
.PRINT AC I(RS) V(7)
.END
```

Table 8.12 shows the PSPICE results. The latter are also available in file ex8_10ps.dat. MATLAB® is used to calculate the input resistance, voltage gain, and also to plot the results.

MATLAB Script

```
% Load data
load 'ex8_10ps.dat' -ascii;
rf = ex8_10ps(:,1);
ib_ac = ex8_10ps(:,2);
vo_ac = ex8_10ps(:,3);
vin_ac = 1.0e-3; % vs is 1 mA
% Calculate the input resistance and gain
n = length(rf); % data points in rf
for i = 1:n
  rin(i) = vin_ac/ib_ac(i);
  gain(i) = vo_ac(i)/vin_ac;
end
%
% Plot input resistance and gain
%
subplot(211)
plot(rf, rin, rf, rin,'ob')
title('(a) Input Resistance versus Feedback Resistance')
ylabel('Input Resistance, Ohms')
subplot(212)
plot(rf, gain,rf,gain,'ob')
title('(b) Amplifier Gain versus Feedback Resistance')
ylabel('Gain')
xlabel('Feedback Resistance, Ohms')
```

The input resistance and gain as a function of feedback resistance are shown in Figure 8.30.

TABLE 8.12

Input Current and Output Voltage as a Function of Feedback Resistance

Feedback Resistance, Ω	Input Current (AC), A	Output Voltage (AC), V
1000	5.430E−06	5.169E−03
2000	5.208E−06	1.001E−02
3000	5.004E−06	1.445E−02
4000	4.817E−06	1.852E−02
5000	4.644E−06	2.228E−02
6000	4.484E−06	2.576E−02
7000	4.336E−06	2.898E−02
8000	4.198E−06	3.197E−02
9000	4.070E−06	3.477E−02
10000	3.950E−06	3.738E−02

(a)

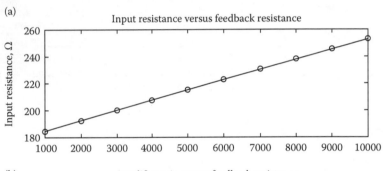

(b)

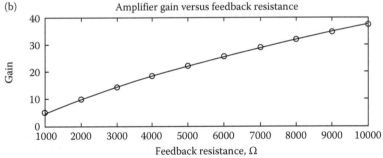

FIGURE 8.30

(a) Input resistance and (b) Gain as a function of feedback resistance.

Problems

8.1   A circuit for determining the input characteristics of a BJT is shown in Figure P8.1. R1 = 1 Ω, R2 = 1 Ω, VCC = 10 V, and Q1 is Q2N2222. (a) Obtain the input characteristics (i.e., IB versus V_{BE}) by varying IB from 1 μA to 9 μA in steps of 2 μA. (b) Obtain the input resistance R_{BE} as a function of I_B.

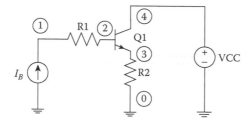

FIGURE P8.1

BJT circuit.

8.2 In Example 8.1, plot the output resistance, r_{CE}, as a function of V_{CE} for $I_B = 4$ μA and 6 μA. Is the output resistance dependent on I_B?

8.3 Figure 8.4 shows a configuration for obtaining MOSFET characteristics. (a) Obtain the output characteristics of MOSFET M2N4531 (i.e., I_{DS} versus V_{DS}) for the following values of V_G: 3, 4, and 5 V. (b) For $V_{GS} = 4$ V, obtain the resistance, $r_{CE} = \Delta V_{DS}/\Delta I_{DS}$ for various values of V_{DS}. Plot r_{CE} versus V_{DS}.

8.4 The data shown in Table P8.4 was obtained from a MOSFET. (a) Determine the threshold voltage, V_T. (b) Determine the transconductance at $V_{GS} = 3$ V.

TABLE P8.4

I_{DS} versus V_{GS} of a MOSFET

V_{GS}, V	I_{DS}, mA
1.0	3.375E−05
2.0	3.375E−05
2.8	3.375E−05
3.0	4.397E−02
3.2	2.093E−01
3.6	7.286E−01
4.0	7.385E−01
4.4	7.418E−01
4.8	7.436E−01

8.5 For the Widlar current source, shown in Figure 8.9, $R_C = 20$ KΩ, $V_{CC} = 5$ V, $R_E = 12$ KΩ. Determine the current I_o as a function of temperature (0°C to 120°C). Assume that the model for both Q1 and Q2 is:

```
.MODEL  npn_mod      NPN(Bf − 150   Br = 2.0   VAF = 125V
Is = 14fA      Tf = 0.35   ns      Rb = 150     + Rc = 150    Re = 2
cje = 1.0 pF      Vje = 0.7 V      Mje = 0.33 Cjc = 0.3 pF Vjc = 0.55
V      mjc = 0.5  + Cjs = 3.0 pF      Vjs = 0.52 V Mjs = 0.5)
```

For resistances, assume that TC1 = 500 ppm/°C and T2 = 0.

8.6   In Example 8.3, determine the worst-case (for all devices) emitter current.

8.7   In Example 8.4, if Q1 is changed to Q2N2222 (model is available in PSPICE device library), and the temperature is varied from −25°C to 55°C, (a) Plot emitter current versus temperature, (b) Determine the best fit between the emitter current and temperature.

8.8   For the MOSFET biasing circuit of Figure 8.14, determine the drain current as R_G takes the following values: 10^4, 10^5, 10^6, 10^7, 10^8, and 10^9 Ω. Assume that $V_{DD} = 15$ V, $R_D = 10$ KΩ, and transistor M1 is 1RF150 (model is available in PSPICE device library).

8.9   In Example 8.7, obtain the input resistance as a function of power supply voltage (7 V to 10 V) when the frequency of the source is 5000 Hz.

8.10 In the common-emitter amplifier shown in Figure 8.18, $C_{C1} = C_{C2} = 5$ μF, $C_E = 100$ μF, $R_{B1} = 50$ KΩ, $R_{B2} = 40$ KΩ, $R_S = 50$ Ω, $R_L = R_C = 10$ KΩ, and $R_E = 1$ KΩ. Transistor Q1 is Q2N3904. Determine (a) gain, (b) input resistance, (c) low cut-off frequency, and (d) bandwidth as a function of supply voltage VCC (8 V to 12 V).

8.11 In Example 8.8, obtain the voltage gain as a function of the emitter resistance.

8.12 The circuit shown in Figure P8.12 is a Darlington amplifier. Q1 and Q2 are the Darlington-pair. The circuit has a very high input resistance. RB = 80 KΩ, RS = 100 Ω, C1 = 5 μF, VCC = 15 V, and transistors Q1 and Q2 are both Q2N2222. If RE varies from 500 Ω to 1500 Ω, determine the input resistance R_{IN} and voltage gain as a function of emitter resistance. Assume that input voltage V_S is sinusoidal waveform with a frequency of 2 KHz and a peak value of 10 mV.

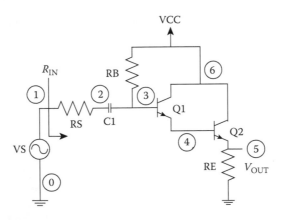

FIGURE P8.12
Darlington amplifier.

8.13 Figure P8.13 is an op amp circuit with series-shunt feedback network. RS = 1 KΩ, RL = 10 KΩ, and R1 = 5 KΩ. Find the gain, V_o/V_S if RF varies from 10 KΩ to 100 KΩ. Plot voltage gain with respect to RF. Assume that the Op Amp is UA741 and the input voltage VS is sinusoidal waveform with a frequency of 5 KHz and a peak voltage of 1 mV.

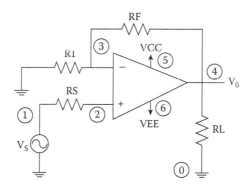

FIGURE P8.13
Op amp circuit with series-shunt feedback network.

8.14 A two-stage amplifier with shunt-series feedback is shown in Figure P8.14. RB1 = 60 KΩ, RB2 = 80 KΩ, RS = 100 Ω, RC1 = 8 KΩ, RE1 = 2.5 KΩ, RB3 = 50 KΩ, RB4 = 60 KΩ, RC2 = 5 KΩ, RE2 = 1 KΩ, C1 = 20 μF, CE = 100 μF, C2 = 20 μF, and VCC = 15 V. If input voltage VS is a sinusoidal voltage with a peak value of 1 mV and a frequency of 1 KHz, determine the input resistance R_{IN} and output voltage as RF varies from 1 KΩ to 6 KΩ. Assume that both transistors Q1 and Q2 are Q2N3904.

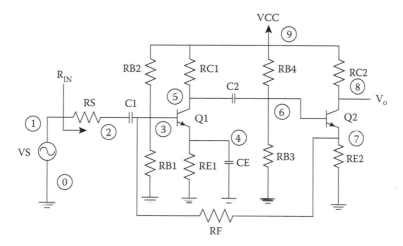

FIGURE P8.14
Two-stage amplifier with shunt-series feedback.

8.15   For Problem 8.14, determine the low cut-off frequency, high cut-off frequency, and the bandwidth as the function of the feedback resistance RF.

8.16   For the two-stage amplifier with shunt-series feedback is shown in Figure P8.14. RB1 = 60 KΩ, RB2 = 80 KΩ, RS = 100 Ω, RC1 = 8 KΩ, RE1 = 2.5 KΩ, RB3 = 50 KΩ, RB4 = 60 KΩ, RC2 = 5 KΩ, RE2 = 1 KΩ, RF = 2 KΩ, C1 = 20μF, CE = 100 μF, and C2 = 20 μF. If input voltage VS is a sinusoidal voltage with a peak value of 1 mV, determine the voltage gain and bandwidth as a function of the supply voltage VCC (8 V to 12 V). Assume that both transistors Q1 and Q2 are Q2N3904.

8.17   For Example 8.10 if VS = 1 mV and frequency of the source is 5 KHz, find the input resistance as R_F varies from 1 KΩ to 8 KΩ.

8.18   In Figure P8.13, if RS = 1 KΩ, RL = 10 KΩ, RF = 20 KΩ, and $V_{CC} = |V_{EE}|$ = 15 V. Find the gain, V_o/V_S if R1 varies from 1 KΩ to 10 KΩ. Plot voltage gain with respect to R1. Assume that the Op Amp is UA741 and the input voltage VS is sinusoidal waveform with a frequency of 2 KHz and a peak voltage of 1 mV.

8.19   For the Darlington amplifier shown Figure P8.12, RB = 60 KΩ, RS = 90 Ω, RE = 1000 Ω, C1 = 10 μF, and transistors Q1 and Q2 are both Q2N2222. If VCC varies from 10 V to 15 V, determine the voltage gain as a function of VCC. Assume that input voltage VS is sinusoidal waveform with a frequency of 1 KHz and a peak value of 5 mV.

8.20   For the common-emitter biasing network, shown in Figure 8.6, V_{CC} = 10 V, R_{B1} = 60 KΩ, R_{B2} = 40 KΩ, R_E = 1 KΩ, R_C = 6 KΩ, and Q1 is Q2N2222. (a) Find the sensitivity of the voltage at the collector to amplifier components.

Bibliography

1. Alexander, Charles K., and Matthew N. O. Sadiku. *Fundamentals of Electric Circuits*. 4th ed. New York: McGraw Hill, 2009.
2. Attia, J. O. *Electronics and Circuit Analysis Using MATLAB®*. 2nd ed. Boca Raton, FL: CRC Press, 2004.
3. Boyd, Robert R. *Tolerance Analysis of Electronic Circuits Using MATLAB®*. Boca Raton, FL: CRC Press, 1999.
4. Chapman, S. J. *MATLAB® Programming for Engineers*. Tampa, FL: Thompson, 2005.
5. Davis, Timothy A., and K. Sigmor. *MATLAB® Primer*. Boca Raton, FL: Chapman & Hall/CRC, 2005.
6. Distler, R. J. "Monte Carlo Analysis of System Tolerance." *IEEE Transactions on Education* 20 (May 1997): 98–101.
7. Etter, D. M. *Engineering Problem Solving with MATLAB®*. 2nd ed. Upper Saddle River, NJ: Prentice Hall, 1997.
8. Etter, D. M., D. C. Kuncicky, and D. Hull. *Introduction to MATLAB® 6*. Upper Saddle River, NJ: Prentice Hall, 2002.

9. Hamann, J. C, J. W. Pierre, S. F. Legowski, and F. M. Long. "Using Monte Carlo Simulations to Introduce Tolerance Design to Undergraduates." *IEEE Transactions on Education* 42, no. 1 (February 1999): 1–14.

10. Gilat, Amos. *MATLAB®, An Introduction With Applications.* 2nd ed. New York: John Wiley & Sons, Inc., 2005.

11 Hahn, Brian D., and Daniel T. Valentine. *Essential MATLAB® for Engineers and Scientists.* 3rd ed. New York and London: Elsevier, 2007.

12. Herniter, Marc E. *Programming in MATLAB®.* Florence, KY: Brooks/Cole Thompson Learning, 2001.

13. Howe, Roger T., and Charles G. Sodini. *Microelectronics, An Integrated Approach.* Upper Saddle River, NJ: Prentice Hall, 1997.

14. Moore, Holly. *MATLAB® for Engineers.* Upper Saddle River, NJ: Pearson Prentice Hall, 2007.

15. Nilsson, James W., and Susan A. Riedel. *Introduction to PSPICE Manual Using ORCAD Release 9.2 to Accompany Electric Circuits.* Upper Saddle River, NJ: Pearson/Prentice Hall, 2005.

16. *OrCAD Family Release 9.2.* San Jose, CA: Cadence Design Systems, 1986–1999.

17. Rashid, Mohammad H. *Introduction to PSPICE Using OrCAD for Circuits and Electronics.* Upper Saddle River, NJ: Pearson/Prentice Hall, 2004.

18. Sedra, A. S., and K. C. Smith. *Microelectronic Circuits.* 5th ed. Oxford: Oxford University Press, 2004.

19. Spence, Robert, and Randeep S. Soin. *Tolerance Design of Electronic Circuits.* London: Imperial College Press, 1997.

20. Soda, Kenneth J. "Flattening the Learning Curve for ORCAD-CADENCE PSPICE," *Computers in Education Journal* XIV (April–June 2004): 24–36.

21. Svoboda, James A. *PSPICE for Linear Circuits.* 2nd ed. New York: John Wiley & Sons, Inc., 2007.

22. Tobin, Paul. "The Role of PSPICE in the Engineering Teaching Environment." Proceedings of International Conference on Engineering Education, Coimbra, Portugal, September 3–7, 2007.

23. Tobin, Paul. *PSPICE for Circuit Theory and Electronic Devices.* San Jose, CA: Morgan & Claypool Publishers, 2007.

24. Tront, Joseph G. *PSPICE for Basic Circuit Analysis.* New York: McGraw-Hill, 2004.

25. *Using MATLAB®, The Language of Technical Computing, Computation, Visualization, Programming, Version 6.* Natick, MA: MathWorks, Inc., 2000.

26. Yang, Won Y., and Seung C. Lee. *Circuit Systems with MATLAB® and PSPICE.* New York: John Wiley & Sons, Inc., 2007.

Index

Printed and bound by CPI Group (UK) Ltd, Croydon, CR0 4YY

21/10/2024

01777089-0013